Mathematical Problems in Relativity, Gravitation, and Cosmology

Physics Research and Technology

More information about this series can be found at
https://novapublishers.com/product-category/series/physics-research-and-technology/

Theoretical and Applied Mathematics

More information about this series can be found at
https://novapublishers.com/product-category/series/theoretical-and-applied-mathematics/

Valeriy Dvoeglazov
Editor

Mathematical Problems in Relativity, Gravitation, and Cosmology

DOI: https://doi.org/10.52305/XLVR1068

NOTICE TO THE READER

The Publisher has taken reasonable care in the preparation of this book, but makes no expressed or implied warranty of any kind and assumes no responsibility for any errors or omissions. No liability is assumed for incidental or consequential damages in connection with or arising out of information contained in this book. The Publisher shall not be liable for any special, consequential, or exemplary damages resulting, in whole or in part, from the readers' use of, or reliance upon, this material. Any parts of this book based on government reports are so indicated and copyright is claimed for those parts to the extent applicable to compilations of such works.

Independent verification should be sought for any data, advice or recommendations contained in this book. In addition, no responsibility is assumed by the Publisher for any injury and/or damage to persons or property arising from any methods, products, instructions, ideas or otherwise contained in this publication.

The Publisher assumes no responsibility for any statements of fact or opinion expressed in the published contents.

This publication is designed to provide accurate and authoritative information with regard to the subject matter covered herein. It is sold with the clear understanding that the Publisher is not engaged in rendering legal or any other professional services. If legal or any other expert assistance is required, the services of a competent person should be sought. FROM A DECLARATION OF PARTICIPANTS JOINTLY ADOPTED BY A COMMITTEE OF THE AMERICAN BAR ASSOCIATION AND A COMMITTEE OF PUBLISHERS.

Additional color graphics may be available in the e-book version of this book.

Library of Congress Cataloging-in-Publication Data

ISBN: 979-8-89530-622-2 (Softcover)
ISBN: 979-8-89530-811-0 (eBook)

Published by Nova Science Publishers, Inc. † New York

Contents

Introduction

This book can be regarded as a continuation of our previously published book series, "Contemporary Fundamental Physics." The thematic volume, "Mathematical Problems in Relativity, Gravitation, Cosmology," includes articles addressing key mathematical challenges in modern physics. This collection may be particularly valuable to unorthodox readers, as unfortunately, several journals have recently ceased publication, including Apeiron, Electronic J. Theor. Phys., Electro-magnetic Phenomena, ICFAI Journal of Physics, and Spacetime and Substance, among others.

The book comprises an Editorial Introduction and eight chapters contributed by renowned researchers. V. Dvoeglazov presents a paper dedicated, as always, to fundamental questions in high-energy physics and gravitation. His recent research was presented at numerous events, including the XII and XIII Workshops (2017, 2019), XI, XII, and XIV Schools (2016, 2018, 2023) of the DGFM of the Sociedad Mexicana de Fisica, the XX Mexican School of Particles and Fields (2023), LXII Congreso Nacional de Fisica (2019), and several international conferences held in Italy, Czechia, Brazil, México, and elsewhere.

V. Abramov explores a ternary algebra involving third-order hypermatrices, introducing "a generalization of the Dirac operator and a ternary generalization of the Pauli exclusion principle." V. Dvoeglazov further addresses negative-energy solutions in relativistic quantum equations.

A. Ignatiev examines "a long-standing quantum mechanical puzzle: whether the collapse of the wave function is a real physical process or simply an epiphenomenon." This puzzle lies at the core of the measurement problem, where "the interaction with apparatus is expected to dominate the measurement process." A valuable discussion and comprehensive bibliography are provided in the second part of his paper.

B. P. Kosyakov (from the Russian Federal Nuclear Center) tackles pedagogical issues in teaching Relativity. His excellent essay honors the 160th anniversary of H. Minkowski's birth, who proposed four-dimensional space in

1904 after earlier attempts. Although many issues covered are well known, this presentation remains valuable for physics students. Indeed, the relation $|x| = ct$ connects space and time, Ref. [1]. Kosyakov also considers the "overdetermined" system of partial differential equations in Maxwell electrodynamics. (cf. Ref. [2]).

Additionally, he notes, "To introduce electric field strength and magnetic induction, Jackson refers to the Lorentz force equation" (though see Dadhich's discussion on monopoles [3]; monopoles remain hypothetical). Unfortunately, the classical "4/3 problem" was not discussed. Next, Kosyakov discussed the relevance of the conformal group $C(1, 3)$ in Relativity. "Maxwell's equations are invariant not only under the 10-parameter Poincaré group of Lorentz transformations, but under the larger, 15-parameter conformal group of spacetime transformations $C(1, 3)$."

Next, "Rainich showed that Maxwell's source-free equations are invariant under... duality rotation," and "there exists a unique nonlinear extension of Maxwell electrodynamics (ModMax theory) that exhibits these maximum allowable symmetries." "ModMax theory has the maximum allowable symmetries". Then what about the action-at-a-distance? Kosyakov states that it "is successfully employed in description of all classical phenomena of electromagnetism and significant part of quantum phenomena". "It is likely that the only opportunity to refute the action-at-a-distance paradigm and confirm the reality of the electromagnetic field is related to the discovery of a magnetic monopole", "...action-at-a-distance paradigm is for optional use".

Again , Kosyakov discussed the absurdity: "A sequence of events of this kind [VD: "faster-than light particles"] is called *causal cycle*". The discussion of the sign of metrics should be compared with Ref. [4]. The philosophical definitions, which follow, can cause a lot of discussions. Particularly, the next statement is interesting: "Poincaré believed that the choice of geometry is a matter of convention. There is no true geometry of our physical world... This idea [gives] rise to attempts to construct theories of gravity in Minkowski space."

A. Krasulin revisits the concepts of bivector gauge fields and bivector derivatives, noting that "the deviation from Coulomb's law considered is a *macroscopic* effect and genuinely represents a deviation from classical electrodynamics."

S. Kruglov again presents his perspective on non-linear electrodynamics. He is motivated by shortcomings of the Standard Cosmological Model (SCM), specifically its failure to resolve critical issues such as initial singularities and cosmic acceleration. Previously proposed scenarios for Universe inflation have included "quintessence (scalar fields)," but Kruglov opts to employ "non-linear electrodynamics (NED) as the source in Einstein's gravity," requiring background stochastic magnetic fields. Readers are encouraged to determine if this approach successfully addresses the outlined objectives.

V. Varlamov begins his discussion with Dirac's observation regarding the limited necessity of classical models in quantum theory, noting importantly that "quantum systems are nonseparable." In my opinion, this mathematically oriented paper raises more questions than it answers, particularly given that many authors emphasize the foundational role of two-component spinors.

Finally, R. Yamaleev employs polynomial representations for dynamical variables, demonstrating that "the planar Hamilton operator of the extended Pauli equation can be represented as a sum of 2s × 2s matrices in degree 2s + 1."

Acknowledgements. And, again about the newest history as in Ref. [5]. The following professors have left our University: Dr. Cuauhtémoc Araujo Andrade (changed university in 2019), Dr. Andrey E. Chubykalo (died in 2020), Dr. D. A. Contreras Solorio (LUMAT-UAZ, retired), Dr. A. Enciso Muños (LUMAT-UAZ, retired), Dr. Miguel Ángel García Aspeitia (changed university in 2021), Dr. A. Espinosa Garrido (LUMAT-UAZ, retired), Dr. R. Ivanov Tsonchev, Dr. A. Pyshchev (UAM-UAZ, changed university, 2024), Dr. Sinhué Lizandro Hinojosa Ruiz (retired, 2022). They were deeply involved in the research at the UAZ School of Physics.

The administrators of the Universidad Autónoma de Zacatecas were: Dr. J. J. Araiza Ibarra (Director of the UAF UAZ, 2008-2012, 2019-2021; Head of Basic Science Studies, 2012-2016), Dr. Sinhué L. Hinojosa Ruiz (Head of the Licenciatura of the UAF-UAZ, 2008-2012, 2019-2020; Director, 2012-2016), Dr. Felipe Román Puch (Director, 2016-2019), Dr. Hugo Tototzintle Huitle (Director, 2021-present), Dr. J. J. Ortega Sigala (Head of the Licenciatura, 2021-present), Dr. C. A. Ortiz González (Head of the Maestría, 2021-present).

Dr. Rubén Ibarra Reyes (General Secretary, 2016-2021; Rector of the UAZ, 2021-present), Dr. Antonio Guzmán Fernández (Rector of the UAZ, 2016-2021), Lic. Miguel Ángel Sánchez Salas (Coordinator of Personel, 2016-2023), Lic. Guadalupe de los Ángeles Escobedo Martínez (Treasurer, 2021-present), Ing. José Juan Martínez Pardo (General Secretary of the SPAUAZ, 2021-2023), Dra. Samanta Desiré Bernal Ayala (Departamento de Servicios Escolares, 2016-present), Lic. Abraham López de Santiago (Defensoría Universitaria), Dr. Juan Manuel Rodríguez Valadez (Defensoría Universitaria).

Unfortunately, the criminal cases (mentioned in Ref. [5]) continued in the UAF-UAZ during last semesters. The students of Mecanica Clásica-I and Física Moderna (jan.-jun. 2023), FM and MC-I (aug.-dec. 2022), Calculo Elemental and FM (aug.-dec. 2020), FM and MC-I (aug.-dec. 2018) obtained "qualifications" by chantage of the professors and by falsifications of the Qualification Acts. Some students are afraid of hongweibing because of bullying. Some professors used them (and even organized them) for political purposes at the University. They impulse students to fight against professors. This is the same as what happened in 1992-93 years when they were students. I hope that future university generations will respect academic freedom ("libertad de catedra"), academic rules and Mexican laws. Otherwise, irresponsibility and lack of education will continue here. (The high caliber of students from 1996 to 2010 was largely due to the presence of several outstanding scientists teaching at UAF-UAZ during that period. For example, several of our students won the prestigious Lederman Prizes and the Bahcall Prize in México.)

Next, the following companies showed low levels of honesty: Aeromexico, Bancomer-BBVA, Banorte, Coppel, Cablecom, DGFM SMF, Estafeta, JIAPAZ, Hospital "San Agustin", Hotel "Don Miguel", Hotel "Paraiso Caxcan", Hotel "Iberostar", KFC, Lasec Technology Systems, Localtel, Movistar-Megacable, Omnibus de México, SPAUAZ, Telcel, tripadvisor.com, Ubereats México, Western Union, Vips, "Me Voy" store, *etc.* We observed some cases of racism, envy and xenofobia, which are not new in our world, including at some Conferences and Journals (e.g., SUSY - de Boer, Ann. Phys. -Finn).

This book can be viewed as a continuation of our previous works in the Book Series. I am grateful to Mrs. N. Columbus (Hauppauge, NY, USA), President of Nova Science Publishers Inc., as well as to Tricia Worthington and Carra Feagaiga, for their invaluable cooperation over the past 25 years. I also sincerely thank Roy Keys for his assistance with the English translation.

Professor Dr. Valeriy V. Dvoeglazov
Universidad Autónoma de Zacatecas
Zacatecas Zac., México

References

[1] V. Acosta et al., Curso de Física Moderna (Harla, México, 1975).
[2] L. M. Hively and A. S. Loebl, Phys. Essays 32 112 (2019).
[3] P. Singh and N. Dadhich, *Mod. Phys. Lett.* A16 83 (2001); *Int. J. Mod. Phys.* A16 1237 (2001).
[4] M. Berg, C. DeWitt-Morrette et al., in *Photon and Poincaré Group*. (Nova Science Pubs., Commack, NY, USA, 1999), p. 83-86.
[5] V. V. Dvoeglazov and A. Enciso Muñoz, *Hadronic J. Suppl.* 17 (2002) – *Proceedings of the Zacatecas School on Theoretical Physics*; V. V. Dvoeglazov, *Einstein and Hilbert: Dark Matter*. (Nova Science Pubs, NY, 2011); V. V. Dvoeglazov, *Einstein and Others: Unification*. (Nova Science Pubs, NY, 2014); *Relativity, Gravitation, Cosmology: Foundations*. (Nova Science Pubs, NY, 2016); *Relativity, Gravitation, Cosmology: Beyond Foundations*. (Nova Science Pubs, NY, 2019); V. V. Dvoeglazov et al., *Future Relativity, Gravitation, Cosmology* (Nova Science Pubs, NY, 2023).

Chapter 1

Biunits of Ternary Algebra of Hypermatrices

Viktor Abramov *
Institute of Mathematics and Statistics, University of Tartu, Estonia

Abstract

We study a ternary algebra of third-order hypermatrices, where by hypermatrix we mean a complex-valued quantity with three indices T_{ijk}. Ternary multiplications of hypermatrices have the property of generalized associativity. We introduce the concepts of q-cyclic and $\bar{q}$-cyclic traceless hypermatrix, where q is a primitive third-order root of unity and study the structures of the corresponding subspaces. We show that q-cyclic traceless hypermatrices can be used to construct right biunits of the ternary algebra of third-order hypermatrices. We show the connection between the Clifford algebra structure induced by the quadratic invariant of q-cyclic traceless hypermatrices and ternary multiplication of hypermatrices. The motivation for studying the space of q-cyclic traceless hypermatrices is the ternary generalization of the Pauli exclusion principle.

Keywords: quark model, Pauli exclusion principle, semi-heap, ternary algebra, hypermatrices, irreducible representation

AMS Subject Classification: 17A40, 20N10, 53C07

---

*Corresponding Author's Email: viktor.abramov@ut.ee

In: Mathematical Problems in Relativity, Gravitation, and Cosmology
Editor: Valeriy Dvoeglazov
ISBN: 979-8-89530-622-2

1. Introduction

In this chapter we study a ternary algebra of hypermatrices. By hypermatrix we mean a quantity with three indices T_{ijk}, where every index is an integer from 1 to n. It should be noted that in this case there is no firmly established terminology, despite the fact that the use of hypermatrices in various fields such as algebra, geometry, theoretical physics and cybernetics is becoming increasingly popular. Quantities with three indices are also called 3-dimensional matrices, spatial matrices or three-index matrices. If we assume the tensor nature of the transformation of such a quantity with three indices, then we get a concept that also has several names, such as trivalent tensor, tensor of rank three or trilinear tensor. In this chapter we will use the terminology proposed by Zapata-Carratalá et al. [18], where the authors advocate the need to establish a unified terminology in this area. Thus, we will call a quantity with three indices T_{ijk} a hypermatrix. The tensor nature of such a quantity is also important for us. A tensor nature will be assured by assuming a naturally defined action of the rotation group $\mathrm{SO}(3)$ on the vector space of hypermatrices.

A vector space of hypermatrices is of interest to abstract algebra primarily because it is a natural, so to speak, building material for ternary multiplication laws. It should be noted that in theoretical physics there is a strong interest in algebras with a ternary multiplication law. Here we note only few approaches. In 1980s an approach to fundamental constituents of matter based on ternary algebras, which are building blocks of Lie algebras and superalgebras was proposed ([7] and references therein) . In the mid-70s of the last century, Nambu proposed a generalization of Hamiltonian mechanics [13], based on a ternary analogue of the Poisson bracket, called the Nambu-Poisson bracket. This approach was later developed by Takhtajan [15]. In the mid-1980s, Filippov proposed a generalization of Lie algebra based on an n-ary Lie bracket and the Filippov-Jacobi identity. Later, a surge of interest in the above generalizations was due to their applications in M-brane theory [5], [6], [8].

The physical motivation for studying ternary algebras of hypermatrices proposed in this chapter is to explore a generalization of the Dirac operator and a ternary extension of the Pauli exclusion principle, as proposed by Kerner [1], [11], [12]. We will briefly explain the ideas that underlie a ternary generalization of the Pauli exclusion principle. First of all, we note that the motivation for a ternary generalization of the Pauli exclusion principle is due to the properties of the quark model. Recall that quarks are the most fundamental particles of matter currently known. This means that all hadrons are composite particles made of quarks. Moreover, a fermionic baryon is a combination of three quarks, and a meson with integer spin is a quark-antiquark pair. Quarks have several quantum characteristics. One of the quantum characteristics is flavor, and in each generation of quarks (there are three of them), there are two flavors $(u, d), (c, s), (t, b)$. In addition to flavor, each quark has a color, electric charge, isospin, baryon number and other quantum characteristics. Note that all allowed combinations of quarks are colorless. In quantum chromodynamics, quarks are treated as fermions with half-integer spin and in this respect they are subject to the Pauli exclusion principle. By analogy with the proton and neutron, which are regarded as two isospin components of a nucleon doublet, the u-quark and d-quark (first generation) can be viewed as two states of a more general entity. According to quantum chromodynamics, in a stable bound state there is place for two quarks in the same u-state or d-state, but not for three. This suggests a possible generalization of the Pauli exclusion principle, which can be formulated as follows: no three quarks with identically equal quantum characteristics can form a stable configuration that is perceived as a strongly interacting particle.

It is well known that the Pauli exclusion principle leads to the skew-symmetry of a wave function of a quantum fermion system. Here we find algebraic properties of a wave function of a system of particles, for example quarks, which obeys the ternary generalization of the Pauli exclusion principle. Consider a quantum system of three particles. Let $|1>, |2>, \ldots, |n>$ be quantum states of these particles and Ψ be a wave function of this system. The value of a wave function on three basic states $|i>, |j>, |k>$ will be denoted by Ψ_{ijk}. Then according to the ternary generalization of the Pauli exclusion principle for any basic state $|i>$ we must have $\Psi_{iii} = 0$. If we now consider the superposition $|\varrho>$ of three basic states $\xi|i> + \zeta|j> + \eta|k>$, where ξ, ζ, η are complex numbers, then, according to the ternary generalization of the Pauli exclusion principle, a wave function must vanish on this superposition, that is, $\Psi_{\varrho\varrho\varrho} = 0$. Making use of the linearity of a wave function we conclude that the equation $\Psi_{\varrho\varrho\varrho} = 0$ holds if and only if for any three basic states $|i>, |j>, |k>$ a wave function Ψ satisfies

$$\Psi_{ijk} + \Psi_{jki} + \Psi_{kij} + \Psi_{kji} + \Psi_{jik} + \Psi_{ikj} = 0, \tag{1}$$

that is, the sum of the values of a wave function on all permutations of any three basic states is equal to zero.

Equation (1) has two solutions:

- A wave function Ψ_{ijk} is totally skew-symmetric, that is, permuting any two basic states in Ψ_{ijk} we get minus. For example, in the case of first two basic states we have $\Psi_{ijk} = -\Psi_{jik}$. It is clear that this solution is consistent with the Pauli exclusion principle, that is, Fermi-Dirac statistics. Note that in this case a wave function vanishes whenever there are two equal states among $|i>, |j>, |k>$. Hence a wave function Ψ_{ijk} is traceless over any pair of basic states, that is,

$$\Psi_{iik} = \Psi_{iki} = \Psi_{kii} = 0. \tag{2}$$

Here and in what follows we use Einstein's convention of summation over repeated indices. Note that in the present case the sum of values of a wave function on cyclic permutations of basic states, that is $\Psi_{ijk} + \Psi_{jki} + \Psi_{kij}$, may be different from zero.

- A wave function Ψ_{ijk} has the property

$$\Psi_{ijk} + \Psi_{jki} + \Psi_{kij} = 0. \tag{3}$$

In this case, a wave function can be nonzero even if we have two identical basic states (with equal quantum characteristics) among $|i >, |j >, |k >$. Thus, this case differs from Fermi-Dirac statistics because a quantum system allows two (but not three!) identical states. Note that, in turn, equation (3) can be also solved if we assume the following properties of the wave function

$$\Psi_{ijk} = q\,\Psi_{jki}, \quad \text{or} \quad \Psi_{ijk} = \bar{q}\,\Psi_{jki},$$

where $q = \exp(2\pi\, i/3)$ is the third-order root of unity. Generally it does not follow from the property (3) that a wave function is traceless over any pair of indices. Therefore, in this part we have a difference from the first solution, that is, the skew-symmetric case. To make the analogy with the skew-symmetric case stronger, we will require that, in addition to the property (3), a wave function must be traceless over any pair of indices, i.e. (2).

In this chapter we study the space of third-order hypermatrices $\mathfrak{T}^3$. A group of rotations $\mathrm{SO}(3)$ acts on this space and under this action the hypermatrices transform like third-order covariant tensors. The vector space $\mathfrak{T}^3$ can be equipped with a ternary multiplication of hypermatrices, which has the property of generalized associativity. In this regard, recall that a set H is called a semi-heap if it is equipped with a ternary multiplication

$$(a, b, c) \in H \times H \times H \to a \cdot b \cdot c \in H,$$

which has the property of generalized associativity

$$(a \cdot b \cdot c) \cdot f \cdot g = a \cdot (f \cdot c \cdot b) \cdot g = a \cdot b \cdot (c \cdot f \cdot g). \tag{4}$$

A semi-heap H is said to be a ternary algebra if H is a vector space. It should be noted that when the bracket in (4) is moved from the leftmost position to the center, elements b and f are rearranged. An excellent overview of the theory of semiheaps is given in [18] (references therein), where in particular it is noted that the concept of a semiheap was introduced by V.V. Wagner in connection with an algebraic approach to the set of transition functions associated with an atlas of a manifold. We will need a notion which generalizes a concept of neutral element to ternary multiplications. An element e of a semiheap H is said to be a right (left) biunit if for any $a \in H$ it is satisfies $a \cdot e \cdot e = a$ ($e \cdot e \cdot a = a$). If $e \in H$ is a right biunit as well as left biunit then it is referred to as a biunit of a semiheap H. A heap is a semiheap H whose any element a is a biunit, that is, for any $b \in H$ it holds $a \cdot a \cdot b = b \cdot a \cdot a = b$.

In order to define a ternary algebra structure on the vector space of third-order hypermatrices we will use the following theorem [2]

Theorem 1. *Let A, B, C be Nth order complex hypermatrices. Then there are only four different triple products of Nth order complex hypermatrices which obey the generalized associativity (4). These are*

1) $(A \odot B \odot C)_{ijk} = A_{ilm} B_{nlm} C_{njk}$, $A \odot B \odot C \to$ A●○○ B○○○ C○●●

2) $(A \odot B \odot C)_{ijk} = A_{ilm} B_{nml} C_{njk}$, $A \odot B \odot C \to$ A●○○ B○○○ C○●●

3) $(A \odot B \odot C)_{ijk} = A_{ijl} B_{nml} C_{mnk}$, $A \odot B \odot C \to$ A●●○ B○○○ C○○●

4) $(A \odot B \odot C)_{ijk} = A_{ijl} B_{mnl} C_{mnk}$, $A \odot B \odot C \to$ A●●○ B○○○ C○○●

This theorem uses diagrams located at the right end of the lines 1), 2), 3), 4) to schematically describe associative ternary multiplications of hypermatrices. In these diagrams, black circles denote free indices (no summation), and summation (or contraction) shown by an arc connecting two white circles is performed over the indices corresponding to white circles. From this theorem it follows that the vector space of Nth order hypermatrices endowed with one of the ternary multiplications 1 - 4 is a ternary algebra. Particularly $\mathfrak{T}^3$ is a ternary algebra. According to the terminology used in [18], the ternary multiplication of 4 is called a fish product. Indeed, ternary multiplication 4 can be schematically represented by the following diagram, which superficially resembles the silhouette of a fish (vertices marked in red indicate summation over the corresponding indices). To differentiate between the ternary multiplications 3 and 4, we will denote them as follows

$$(T \diamond U \diamond V)_{ijk} = T_{ijp} U_{rsp} V_{srk}, \quad (T \bullet U \bullet V)_{ijk} = T_{ijp} U_{rsp} V_{rsk}, \tag{5}$$

where $T, U, V \in \mathfrak{T}^3$. The corresponding ternary algebras of hypermatrices will be denoted as pairs $(\mathfrak{T}^3, \diamond), (\mathfrak{T}^3, \bullet)$. When a statement holds for both ternary multiplications, we will denote them with a single symbol $\odot$.

In the theory of tensor representations of the rotation group [10] it is shown that the space of third-order hypermatrices $\mathfrak{T}^3$ can be uniquely decomposed into a direct sum of subspaces $\mathfrak{T}^3_a$, $a = 0, 1, 2, 3$, where a is a weight of representation. In each

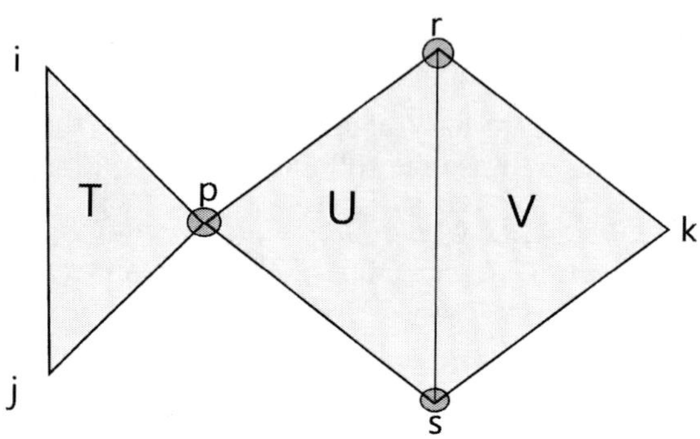

Figure 1. Ternary multiplication $(T \bullet U \bullet V)_{ijk} = T_{ijp} U_{rsp} V_{rsk}$.

of the subspaces $\mathfrak{T}^3_a$ we have a representation of the rotation group multiple to an irreducible one. Of particular interest to us are subspaces $\mathfrak{T}^3_0$ and $\mathfrak{T}^3_2$. The direct sum of these spaces is the space of traceless hypermatrices and $\mathfrak{T}^3_0$ is the subspace of completely skew-symmetric hypermatrices, and $\mathfrak{T}^3_2$ is the subspace of traceless hypermatrices with the property

$$T_{ijk} + T_{jki} + T_{kij} = 0. \tag{6}$$

In $\mathfrak{T}^3_0$ there is an irreducible representation of $\mathrm{SO}(3)$ of the weight zero and in $\mathfrak{T}^3_2$ there is a twofold irreducible representation of $\mathrm{SO}(3)$ of the weight 2. In order to split the twofold irreducible representation into two irreducible representations one can decompose the subspace $\mathfrak{T}^3_2$ into a direct sum of two subspaces by solving the equation (6) as it was shown above (in the case of a wave function)

$$\begin{aligned} T_{ijk} &= q\, T_{jki}, \quad (q - \text{cyclic hypermatrix}), \\ T_{ijk} &= \bar{q}\, T_{jki}, \quad (\bar{q} - \text{cyclic hypermatrix}). \end{aligned} \tag{7}$$

In this chapter, we calculate the quadratic $\mathrm{SO}(3)$-invariants of q-cyclic hypermatrices and the list of these invariants can be found in [4]. We prove a theorem which states that if T is a q-cyclic hypermatrix such that $I_2 \neq 0$, where I_2 is a quadratic $\mathrm{SO}(3)$-invariant, then T multiplied by an appropriate factor is a right biunit of the ternary algebra $(\mathfrak{T}^3, \diamond)$. The quadratic invariant I_2 induces a non-degenerate bilinear form K on the space of q-cyclic hypermatrices. This non-degenerate bilinear form induces a Clifford algebra structure on the complex 5-dimensional space of q-cyclic hypermatrices. We show the connection between Clifford algebra structure and ternary hypermatrix multiplication $T \diamond U \diamond V$.

2. Ternary Algebra of Hypermatrices

In this section we consider the vector space of complex third-order hypermatrices. This vector space will be denoted by $\mathfrak{T}^3$. The dimension of this vector space is 27. The rotation group $\mathrm{SO}(3)$ acts on the vector space of complex third-order hypermatrices as follows

$$\tilde{T}_{prs} = g_{pi}\, g_{rj}\, g_{sk}\, T_{ijk}, \quad g = (g_{ij}) \in \mathrm{SO}(3), \ T, \tilde{T} \in \mathfrak{T}^3. \tag{8}$$

Here and henceforth we use the Einstein's convention of summation over twice repeated indexes. It is worth mentioning that the formula (8) defines the tensor representation of the rotation group $\mathrm{SO}(3)$ in the complex vector space $\mathfrak{T}^3$.

Thus we can identify the components of a third-order hypermatrix T with components of the third-order covariant tensor. We assume that the entries of a hypermatrix T are located in 3-dimensional space in such a way that connecting them with straight line segments we get a cube.

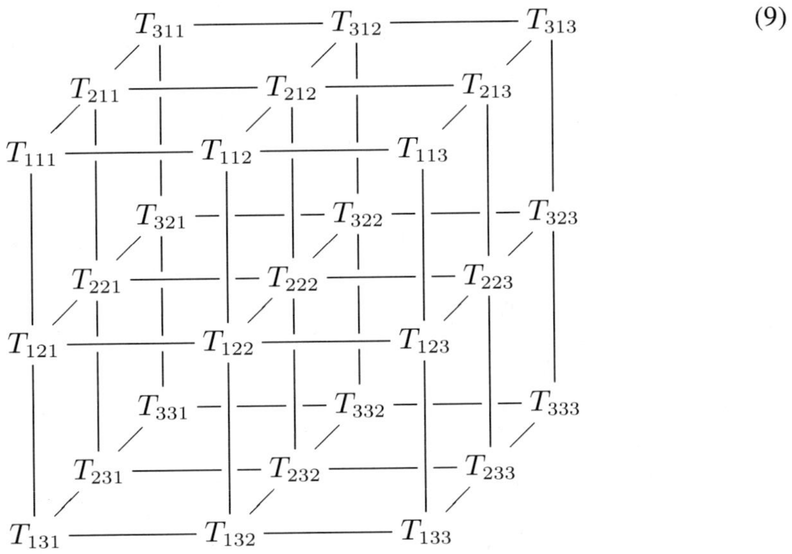

(9)

We can define three directions in a hypermatrix T. The direction of increase of the index i (first index) will be referred to as the i-direction of a hypermatrix T. Similarly, we define the j-direction and k-direction of a hypermatrix. It is useful to split a hypermatrix T into third-order square matrices. This is usually done by sections of the cube of a hypermatrix by planes orthogonal to edges of the cube or to three directions of a hypermatrix defined above. If we cut the cube (9) with three planes perpendicular to the i-direction of hypermatrix we get three third-order square matrices corresponding to the values $i = 1, 2, 3$, which we denote by $T_1^{(1)} = (T_{1jk}), T_2^{(1)} = (T_{2jk}), T_3^{(1)} = (T_{3jk})$ respectively. In other words the subscript (1) shows that we fix the first subscript in T_{ijk} and the subscript shows the value of the fixed subscript. Arranging these three square matrices one after the other from left

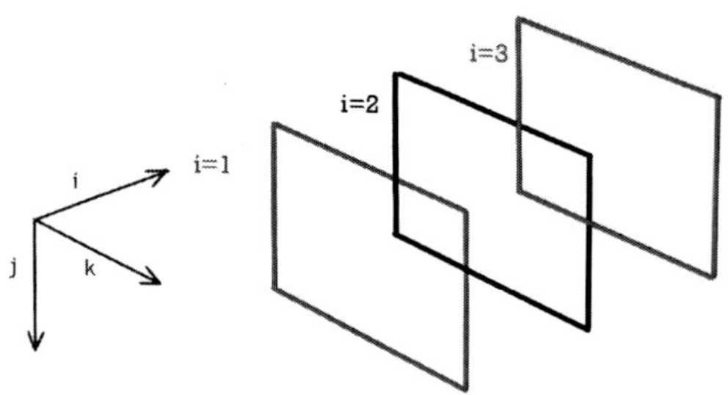

Figure 2. Sections of a hypermatrix perpendicular to i-direction.

to right in the plane of the page and separating them by vertical lines, we obtain a useful notation for the hypermatrix T in the form

$$T = \left(\begin{array}{ccc|ccc|ccc} T_{111} & T_{112} & T_{113} & T_{211} & T_{212} & T_{213} & T_{311} & T_{312} & T_{313} \\ T_{121} & T_{122} & T_{123} & T_{221} & T_{222} & T_{223} & T_{321} & T_{322} & T_{323} \\ T_{131} & T_{132} & T_{133} & T_{231} & T_{232} & T_{233} & T_{331} & T_{332} & T_{333} \end{array} \right). \quad (10)$$

Analogously, fixing the value of the second subscript j (the third subscript k) in T_{ijk} we get the square matrices $T_1^{(2)}, T_2^{(2)}, T_3^{(2)}$ $(T_1^{(k)}, T_2^{(k)}, T_3^{(k)})$. For example, in order to write out the entries of the matrix $T_3^{(3)}$ using hypermatrix (10) we must take the third column in each square matrix $T_1^{(1)}, T_2^{(1)}, T_3^{(1)}$, compose them into a square matrix, and then transpose it.

Such important operations with square matrices as determinant and trace can be extended to the space of hypermatrices by constructing corresponding analogues [14]. In this chapter we will be interested in an algebraic structure based on a ternary multiplication of hypermatrices. Recall that a semiheap is a set H equipped with a ternary multiplication $(a, b, c) \in H \times H \times H \to a \cdot b \cdot c \in H$ which for any five elements $a, b, c, d, u, v \in H$ satisfies the generalized associativity

$$(a \cdot b \cdot c) \cdot u \cdot v = a \cdot (u \cdot c \cdot b) \cdot v = a \cdot b \cdot (c \cdot u \cdot v). \quad (11)$$

In this chapter we will study the case when a set H has the structure of a vector space. In this case, a semiheap H is called a ternary algebra. Thus, ternary algebra is a vector space equipped with an associative (11) ternary multiplication law. A wide class of ternary algebras can be constructed using hypermatrices.

In particular, the complex vector space of third-order tensors $\mathfrak{T}^3$, equipped with one of the ternary multiplications 1 - 4 shown in Theorem 1, is a ternary algebra and this algebra is the main object of study in this chapter. Since the structure of ternary multiplications 1,2 is similar to the structures of ternary multiplications 3,4, we will study ternary algebras of hypermatrices with ternary multiplications 3,4.

3. Invariants and Irreducible Representations

In this chapter we will need invariants of the third-order hypermatrices under the action of the rotation group. The complete set of linear and quadratic SO(3)-invariants is given in [4]. For third-order hypermatrix T there is only one linear SO(3)-invariant

$$I = \epsilon_{ijk}T_{ijk} = T_{123} + T_{231} + T_{312} - T_{321} - T_{213} - T_{132}.$$

The complete set of quadratic SO(3)-invariants of a complex hypermatrix $T = (T_{ijk})$ consists of

$$\begin{aligned}
I_1 &= T_{ijk}T_{ijk},\ I_1^* = T_{ijk}\overline{T}_{ijk},\ I_2 = T_{ijk}T_{ikj},\ I_2^* = T_{ijk}\overline{T}_{ikj},\\
I_3 &= T_{ijk}\,T_{jik},\ I_3^* = T_{ijk}\,\overline{T}_{jik},\ I_4 = T_{ijk}\,T_{kji},\ I_4^* = T_{ijk}\overline{T}_{kji},\\
I_5 &= T_{ijk}T_{kij} + T_{ijk}T_{jki},\ I_5^* = T_{ijk}\overline{T}_{kij} + T_{ijk}\overline{T}_{jki},\\
I_6 &= T_{iik}T_{ppk},\ I_6^* = T_{iik}\overline{T}_{ppk},\ I_7 = T_{iji}T_{pjp},\ I_7^* = T_{iji}\overline{T}_{pjp},\\
I_8 &= T_{ijj}T_{iqq},\ I_8^* = T_{ijj}\overline{T}_{iqq},\ I_9 = T_{iik}T_{kqq},\ I_9^* = T_{iik}\overline{T}_{kqq},\\
I_{10} &= \frac{1}{2}(T_{iik}T_{pkp} + T_{iji}T_{ppj}),\ I_{10}^* = \frac{1}{2}(T_{iik}\overline{T}_{pkp} + T_{iji}\overline{T}_{ppj}),\\
I_{11} &= \frac{1}{2}(T_{iji}T_{jqq} + T_{ijj}T_{pip}),\ I_{11}^* = \frac{1}{2}(T_{iji}\overline{T}_{jqq} + T_{ijj}\overline{T}_{pip}).
\end{aligned} \tag{12}$$

Recall that the rotation group acts on the vector space of hypermatrices $\mathfrak{T}^3$ and this action encodes the tensor nature of the hypermatrices. Thus, any structure on the vector space of hypermatrices must be consistent with the action of the rotation group, that is, be invariant with respect to this action. Let us denote this action as follows

$$(g, T) \in \mathrm{SO}(3) \times \mathfrak{T}^3 \to g \cdot T \in \mathfrak{T}^3. \tag{13}$$

Then (8) can be written as

$$(g \cdot T)_{prs} = g_{pi}\, g_{rj}\, g_{sk}\, T_{ijk}.$$

Since the ternary multiplications (5) are constructed by means of contractions of tensors over a pair of indices, it is obvious that the ternary product of three tensors

T, U, V is an SO(3)-tensor. Hence the ternary multiplications (5) commute with the action of the rotation group (or its tensor representation), that is, we have

$$(g\cdot T)\odot(g\cdot U)\odot(g\cdot V)=g\cdot(T\odot U\odot V).$$

Another SO(3)-invariant structure on the ternary algebra $\mathfrak{T}^3$ is a Hermitian metric. We use the invariant I_1^* (12) to construct a Hermitian metric on the ternary algebra $\mathfrak{T}^3$. Hence we define the Hermitian scalar product of two hypermatrices $T, U\in\mathfrak{T}^3$ as follows

$$h(T,U)=T_{ijk}\,\overline{U}_{ijk}. \tag{14}$$

This way of defining the Hermitian metric h is natural, since in this case we consider the entries of a hypermatrix as the coordinates of the 27-dimensional complex vector. Evidently

$$h(g\cdot T, g\cdot U)=h(T,U).$$

In the theory of representations of the rotation group [10], it is shown that the representation of the rotation group in the complex vector space of hypermatrices of third order (13) can be uniquely decomposed into irreducible representations as follows

$$\mathfrak{T}^3=\oplus_{w=0}^3\,\mathfrak{T}^3_w,$$

where w is a weight of representation, $\mathfrak{T}^3_0$ is the vector space of totally skew-symmetric third-order hypermatrices, $\mathfrak{T}^3_1$ is the vector space of third-order hypermatrices of the type

$$T_{ijk}=\delta_{ij}T_k+\delta_{ik}T_j+\delta_{jk}T_i,$$

$\mathfrak{T}^3_2$ is the vector space of traceless hypermatrices (trace over any pair of indices is zero) satisfying the equation

$$T_{ijk}+T_{jki}+T_{kij}=0,$$

and $\mathfrak{T}^3_3$ is the vector space of traceless totally symmetric hypermatrices. Thus

$$\begin{aligned}
\mathfrak{T}^3_0 &= \{T\in\mathfrak{T}^3: T_{ijk}=\lambda\,\epsilon_{ijk}, \lambda\in\mathbb{C}\},\\
\mathfrak{T}^3_1 &= \{T\in\mathfrak{T}^3: T_{ijk}=\delta_{ij}T_k+\delta_{ik}T_j+\delta_{jk}T_i\},\\
\mathfrak{T}^3_2 &= \{T\in\mathfrak{T}^3: T_{iik}=0,\ T_{iji}=0,\ T_{ijj}=0,\ T_{ijk}+T_{jki}+T_{kij}=0\},\\
\mathfrak{T}^3_3 &= \{T\in\mathfrak{T}^3: T_{iik}=0,\ T_{iji}=0,\ T_{ijj}=0, T_{i_{\sigma(1)}i_{\sigma(2)}i_{\sigma(3)}}=T_{i_1i_2i_3}, \forall\sigma\in S_3\},
\end{aligned}$$

where ϵ_{ijk} is the Levi-Civita symbol and S_3 is the group of substitutions of the set $\{1,2,3\}$. These subspaces have the following dimensions

$$\dim\mathfrak{T}^3_0=1,\ \ \dim\mathfrak{T}^3_1=9,\ \ \dim\mathfrak{T}^3_2=10,\ \ \dim\mathfrak{T}^3_3=7.$$

In each of these subspaces we have a representation of the rotation group that is a multiple of an irreducible one. In the subspace $\mathfrak{T}_0^3$ we have only one irreducible representation (of weight 0), in the subspace $\mathfrak{T}_1^3$ a three times repeated (threefold) irreducible representation (of weight 1), in the subspace $\mathfrak{T}_2^3$ a twice repeated (twofold) irreducible representation (of weight 2), and in the subspace $\mathfrak{T}_3^3$ we have one irreducible representation (of weight 3). It is worth noting that every subspace $\mathfrak{T}_w^3$ is invariant under the action of the rotation group. It is also worth noting that the direct sum of three subspaces $\mathfrak{T}_0^3, \mathfrak{T}_2^3, \mathfrak{T}_3^3$ is the subspace of traceless third-order hypermatrices (trace over any pair of subscripts is zero), which we denote by $\mathfrak{T}_{\mathtt{tr}}^3$. Hence $\mathfrak{T}^3 = \mathfrak{T}_1^3 \oplus \mathfrak{T}_{\mathtt{tr}}^3$, where $\dim \mathfrak{T}_{\mathtt{tr}}^3 = 18$.

In order to decompose a representation that is a multiple of an irreducible representation into irreducible representations, we must decompose the representation space into a direct sum of subspaces (their number is equal to the multiplicity of a representation) so that this decomposition is invariant under the action of the rotation group. In this chapter, we will study the subspace of the weight 2 representation, that is, the subspace $\mathfrak{T}_2^3$. This is because the subspace $\mathfrak{T}_2^3$ is most closely related to the ternary generalization of the Pauli principle. In this subspace there is a twofold irreducible representation of the rotation group. Hence in order to split this twofold irreducible representation into two irreducible representations we have to split the representation space $\mathfrak{T}_2^3$ into a direct sum of two subspaces invariant with respect to the action of the rotation group. For this purpose we will use a substitution operator. Let $\varsigma \in S_3$ be the cyclic substitution $\varsigma(1) = 2, \varsigma(2) = 3, \varsigma(3) = 1$. Define the substitution operator $\mathrm{L}_\varsigma : \mathfrak{T}^3 \to \mathfrak{T}^3$ as follows $\mathrm{L}_\varsigma(T)_{i_1 i_2 i_3} = T_{i_{\varsigma(1)} i_{\varsigma(2)} i_{\varsigma(3)}} = T_{i_2 i_3 i_1}$. Obviously the substitution operator L_ς is a linear operator and $\mathrm{L}_\varsigma^3 = \mathtt{Id}$, where $\mathtt{Id}$ is the identity operator. The latter implies that the substitution operator L_ς has three eigenvalues $1, q, \bar{q}$, where $q = \exp(2i\pi/3)$ is the primitive third order root of unity and $\bar{q}$ is its complex conjugate. It is well known that $1 + q + \bar{q} = 0$. Hence we can decompose the vector space of third-order hypermatrices $\mathfrak{T}^3$ into the direct sum of three subspaces $\mathfrak{T}^{3,1}, \mathfrak{T}^{3,q}, \mathfrak{T}^{3,\bar{q}}$ corresponding to the eigenvalues $1, q, \bar{q}$ respectively. Hence

$$\mathfrak{T}^3 = \mathfrak{T}^{3,1} \oplus \mathfrak{T}^{3,q} \oplus \mathfrak{T}^{3,\bar{q}}, \tag{15}$$

where

$$\begin{aligned} \mathfrak{T}^{3,1} &= \{T \in \mathfrak{T}^3 : \mathrm{L}_\varsigma(T) = T\}, \\ \mathfrak{T}^{3,q} &= \{T \in \mathfrak{T}^3 : \mathrm{L}_\varsigma(T) = q\,T\}, \\ \mathfrak{T}^{3,\bar{q}} &= \{T \in \mathfrak{T}^3 : \mathrm{L}_\varsigma(T) = \bar{q}\,T\}. \end{aligned}$$

The decomposition (15) can be obtained from more elementary considerations. It is easy to see that any third-order hypermatrix T can be represented in the form

$$T_{ijk} = \frac{1}{3}(T_{ijk} + T_{jki} + T_{kij}) + \frac{1}{3}(T_{ijk} + \bar{q}\,T_{jki} + q\,T_{kij}) + \frac{1}{3}(T_{ijk} + q\,T_{jki} + \bar{q}\,T_{kij}).$$

In order to give this decomposition a more concise form we introduce the following square polynomials of the substitution operator

$$\begin{aligned}\xi_1 &= \frac{1}{3}(\mathtt{Id} + \mathrm{L}_\varsigma + \mathrm{L}_\varsigma^2),\\ \xi_q &= \frac{1}{3}(\mathtt{Id} + \bar{q}\,\mathrm{L}_\varsigma + q\,\mathrm{L}_\varsigma^2),\\ \xi_{\bar{q}} &= \frac{1}{3}(\mathtt{Id} + q\,\mathrm{L}_\varsigma + \bar{q}\,\mathrm{L}_\varsigma^2).\end{aligned}$$

Thus, the operators $\xi_1, \xi_q, \xi_{\bar{q}}$ are operators of projection of the hypermatrix onto subspaces $\mathfrak{T}^{3,1}, \mathfrak{T}^{3,q}, \mathfrak{T}^{3,\bar{q}}$ respectively. It is easy to verify the following properties

$$\xi_1 + \xi_q + \xi_{\bar{q}} = \mathtt{Id},\ \mathrm{L}_\xi\,\xi_1 = \xi_1\,\mathrm{L}_\xi = \xi_1,\ \mathrm{L}_\xi\,\xi_q = \xi_q\,\mathrm{L}_\xi = q\,\xi_q, \mathrm{L}_\xi\,\xi_{\bar{q}} = \xi_{\bar{q}}\,\mathrm{L}_\xi \ = \bar{q}\,\xi_{\bar{q}}.$$

Now we can describe the weight 2 representation vector space $\mathfrak{T}_2^3$ as follows

$$\mathfrak{T}_2^3 = \mathrm{Ker}\,\xi_1 \cap \mathfrak{T}^3_{\mathtt{tr}}.$$

It is easy to verify that the subspace $\mathfrak{T}_2^3$ is invariant with respect to the substitution operator L_ς, that is, the restriction of the operator L_ς to the subspace $\mathfrak{T}_2^3$ is a correctly defined linear operator on $\mathfrak{T}_2^3$, i.e. $\mathrm{L}_\varsigma : \mathfrak{T}_2^3 \to \mathfrak{T}_2^3$. Assume that $T \in \mathfrak{T}^3_{\mathtt{tr}}$. Then $\mathrm{L}_\varsigma(T) \in \mathfrak{T}^3_{\mathtt{tr}}$. Indeed

$$\mathrm{L}_\varsigma(T)_{iij} = T_{iji} = 0,$$

and analogously for the other traces. From the property $\mathrm{L}_\xi\,\xi_1 = \xi_1\,\mathrm{L}_\xi = \xi_1$ it follows that the subspace $\mathrm{Ker}\,\xi_1$ is invariant under the action of the operator L_ξ.

Thus the 10-dimensional weight 2 representation vector space splits into two 5-dimensional subspaces $\mathfrak{T}_2^3 = \mathfrak{T}_2^{3,q} \oplus \mathfrak{T}_2^{3,\bar{q}}$, where $\mathfrak{T}_2^{3,q} = \mathfrak{T}^{3,q} \cap \mathfrak{T}_2^3$ and $\mathfrak{T}_2^{3,\bar{q}} = \mathfrak{T}^{3,\bar{q}} \cap \mathfrak{T}_2^3$. It is worth noting that $\mathfrak{T}^{3,1} \cap \mathfrak{T}_2^3 = \{0\}$. Thus in each 5-dimensional subspace $\mathfrak{T}_2^{3,q}, \mathfrak{T}_2^{3,\bar{q}}$ we have an irreducible representation of the rotation group. Since the resulting subspaces $\mathfrak{T}_2^{3,q}, \mathfrak{T}_2^{3,\bar{q}}$ are very important for what follows, it is useful to give them an explicit description

$$\begin{aligned}\mathfrak{T}_2^{3,q} &= \{T \in \mathfrak{T}^3 : T_{iij} = 0,\ T_{iji} = 0,\ T_{jii} = 0,\ T_{ijk} = \bar{q}\,T_{jki}\}, \quad (16)\\ \mathfrak{T}_2^{3,\bar{q}} &= \{T \in \mathfrak{T}^3 : T_{iij} = 0,\ T_{iji} = 0,\ T_{jii} = 0,\ T_{ijk} = q\,T_{jki}\}. \quad (17)\end{aligned}$$

The notations $\mathfrak{T}_2^{3,q}$ and $\mathfrak{T}_2^{3,\bar{q}}$ are quite complex. Unfortunately, we have not found simpler notations that would include all the information about these subspaces. Formulas (16) and (17) show that in fact the hypermatrices of these subspaces are determined by only two conditions, where one is that they are traceless, and the second is a transformation law of hypermatrix entries under a cyclic permutation of subscripts. Therefore, to simplify the presentation, we will call hypermatrices of (16) traceless $\bar{q}$-cyclic hypermatrices, meaning by the latter that $T_{ijk} = \bar{q}\,T_{jki}$. Note that the same formula can be represented in the form $T_{jki} = q\,T_{ijk}$ that clearly shows that such hypermatrices are eigenvectors of the substitution operator L_ς with the eigenvalue q. This is the reason why we use q in the notation $\mathfrak{T}_2^{3,q}$ for the subspace of this kind of hypermatrices. Analogously the hypermatrices of (17) will be referred to as traceless q-cyclic hypermatrices, where q-cyclic stands for $T_{ijk} = q\,T_{jki}$.

It is easy to see that complex conjugation maps subspace $\mathfrak{T}_2^{3,q}$ to subspace $\mathfrak{T}_2^{3,\bar{q}}$ and vice versa. Indeed, if a traceless hypermatrix $T = (T_{ijk})$ transforms under cyclic permutation ς according to the formula $T_{ijk} = \bar{q}\,T_{jki}$, i.e. $T \in \mathfrak{T}_2^{3,q}$, then its complex conjugate $\overline{T} = (\overline{T}_{ijk})$ will transform under the same cyclic permutation according to the formula $\overline{T}_{ijk} = \overline{\bar{q}\,T_{jki}} = q\,\overline{T}_{jki}$, i.e. $\overline{T} \in \mathfrak{T}_2^{3,\bar{q}}$.

4. The Space of Traceless q-cyclic Hypermatrices

Our aim now is to study the structure of the 5-dimensional complex vector space of traceless q-cyclic hypermatrices $\mathfrak{T}_2^{3,\bar{q}}$. In this space we have an irreducible representation of the rotation group SO(3). We choose the following five traceless q-cyclic hypermatrices as a basis

$$
\begin{aligned}
E_1 &= \frac{1}{\sqrt{6}}\left(\begin{array}{ccc|ccc|ccc} 0 & 0 & 0 & 0 & q & 0 & 0 & 0 & -q \\ 0 & 1 & 0 & \bar{q} & 0 & 0 & 0 & 0 & 0 \\ 0 & 0 & -1 & 0 & 0 & 0 & -\bar{q} & 0 & 0 \end{array}\right), \\
E_2 &= \frac{1}{\sqrt{6}}\left(\begin{array}{ccc|ccc|ccc} 0 & -\bar{q} & 0 & -1 & 0 & 0 & 0 & 0 & 0 \\ -q & 0 & 0 & 0 & 0 & 0 & 0 & 0 & q \\ 0 & 0 & 0 & 0 & 0 & 1 & 0 & \bar{q} & 0 \end{array}\right), \\
E_3 &= \frac{1}{\sqrt{6}}\left(\begin{array}{ccc|ccc|ccc} 0 & 0 & \bar{q} & 0 & 0 & 0 & 1 & 0 & 0 \\ 0 & 0 & 0 & 0 & 0 & -\bar{q} & 0 & -1 & 0 \\ q & 0 & 0 & 0 & -q & 0 & 0 & 0 & 0 \end{array}\right), \\
E_4 &= \frac{1}{\sqrt{3}}\left(\begin{array}{ccc|ccc|ccc} 0 & 0 & 0 & 0 & 0 & 0 & 0 & q & 0 \\ 0 & 0 & 1 & 0 & 0 & 0 & 0 & 0 & 0 \\ 0 & 0 & 0 & \bar{q} & 0 & 0 & 0 & 0 & 0 \end{array}\right), \\
E_5 &= \frac{1}{\sqrt{3}}\left(\begin{array}{ccc|ccc|ccc} 0 & 0 & 0 & 0 & 0 & \bar{q} & 0 & 0 & 0 \\ 0 & 0 & 0 & 0 & 0 & 0 & 1 & 0 & 0 \\ 0 & q & 0 & 0 & 0 & 0 & 0 & 0 & 0 \end{array}\right).
\end{aligned}
\tag{18}
$$

It is easy to verify that this basis is an orthonormal basis with respect to the Hermitian metric h (14), that is, $h(E_A, E_B) = \delta_{AB}$, where the subscripts denoted by capital Latin letters run from 1 to 5. Using the orthonormal basis $\{E_A\}$, we can identify a vector $z = (z^A)$ of the 5-dimensional complex vector space $\mathbb{C}^5$ with the third-order hypermatrix $T(z)$ as follows

$$z = (z^A) \in \mathbb{C}^5 \mapsto T(z) = z^A\, E_A \in \mathfrak{T}_2^{3,\bar{q}},$$

where

$$T(z) = \left(\begin{array}{ccc|ccc|ccc} 0 & -\frac{\bar{q}\, z^2}{\sqrt{6}} & \frac{\bar{q}\, z^3}{\sqrt{6}} & -\frac{z^2}{\sqrt{6}} & \frac{q\, z^1}{\sqrt{6}} & \frac{\bar{q}\, z^5}{\sqrt{3}} & \frac{z^3}{\sqrt{6}} & \frac{q\, z^4}{\sqrt{3}} & -\frac{q\, z^1}{\sqrt{6}} \\ -\frac{q\, z^2}{\sqrt{6}} & \frac{z^1}{\sqrt{6}} & \frac{z^4}{\sqrt{3}} & \frac{\bar{q}\, z^1}{\sqrt{6}} & 0 & -\frac{\bar{q}\, z^3}{\sqrt{6}} & \frac{z^5}{\sqrt{3}} & -\frac{z^3}{\sqrt{6}} & \frac{q\, z^2}{\sqrt{6}} \\ \frac{q\, z^3}{\sqrt{6}} & \frac{q\, z^5}{\sqrt{3}} & -\frac{z^1}{\sqrt{6}} & \frac{\bar{q}\, z^4}{\sqrt{3}} & -\frac{q\, z^3}{\sqrt{6}} & \frac{z^2}{\sqrt{6}} & -\frac{\bar{q}\, z^1}{\sqrt{6}} & \frac{\bar{q}\, z^2}{\sqrt{6}} & 0 \end{array}\right).$$

Throughout what follows we will keep in mind that the 5-dimensional complex vector space $\mathbb{C}^5$ is identified with the subspace of third-order hypermatrices $\mathfrak{T}_2^{3,\bar{q}}$. In other words, we consider a 5-dimensional complex (Hermitian) space, each point of which is a third-order hypermatrix (three-dimensional matrix), that is, it has the shape of a cube (9).

Since the subspace $\mathfrak{T}_2^{3,\bar{q}}$ is invariant under the action of the rotation group $\mathrm{SO}(3)$, the set of invariants (12), when restricted to the subspace $\mathfrak{T}_2^{3,\bar{q}}$, will give the set of subspace $\mathfrak{T}_2^{3,\bar{q}}$ invariants. Calculating these invariants in coordinates z^A, we get

$$I_1 = 0, \qquad I_1^* = \sum_{A=1}^{5} z^A\, \bar{z}^A = h(z, \bar{z}), \tag{19}$$

$$I_2 = \sum_{A=1}^{3} (z^A)^2 + 2\, q\, z^4 z^5, \quad I_2^* = 0, \tag{20}$$

$$I_3 = q\, I_2, \qquad I_3^* = \bar{q}\, I_2^* = 0, \tag{21}$$

$$I_4 = \bar{q}\, I_2, \qquad I_4^* = q\, I_2^* = 0, \tag{22}$$

$$I_5 = 0, \qquad I_5^* = -I_1^*, \tag{23}$$

and the invariants $I_6, I_6^*, I_7, I_7^*, I_8, I_8^*, I_9, I_9^*, I_{10}, I_{10}^*, I_{11}, I_{11}^*$ vanish because they are constructed by means of the trace of a hypermatrix (hypermatrices in $\mathfrak{T}_2^{3,\bar{q}}$ are traceless). Thus we have two independent invariants in the subspace $\mathfrak{T}_2^{3,\bar{q}}$, where one is the canonical Hermitian metric $h(z, \bar{z})$ and the other is the quadratic form $(z^1)^2 + (z^2)^2 + (z^3)^2 + 2q\, z^4 z^5$, which will be denoted by $K(z, z)$.

The quadratic invariant $K(z, z)$ will play an important role in what follows. In paper [3] the properties of this invariant were studied. Let $K(z, z) = K_{AB} z^A z^B$, where K_{AB} is the 5th order matrix of the quadratic form $K(z,z)$

$$(K_{AB}) = \begin{pmatrix} 1 & 0 & 0 & 0 & 0 \\ 0 & 1 & 0 & 0 & 0 \\ 0 & 0 & 1 & 0 & 0 \\ 0 & 0 & 0 & 0 & q \\ 0 & 0 & 0 & q & 0 \end{pmatrix}.$$

The matrix (K_{AB}) can also be considered as a twice covariant tensor in the 5-dimensional complex vector space $\mathbb{C}^5$, that is,

$$\tilde{K}_{AB} = \mathrm{U}_A^C\, \mathrm{U}_B^D\, K_{CD}, \tag{24}$$

where (U_A^C) is a 5th order unitary matrix. For any orthonormal basis $\{E_A\}$ the second-order covariant tensor $K_{AB} = K(E_A, E_B)$ determined by the quadratic form $K(z,z)$ has the following properties which are unitary invariant:

- $K_{AB} = K_{BA}$ (symmetric),
- $K_{AB}\,\overline{K}_{CB} = \delta_{AB}$ (unitary),
- $\det(K_{AB}) = \epsilon$, where $\epsilon = e^{i\pi/3}$ is the sixth order root of unity.

It also has the following properties, which are invariant with respect to real unitary transformations:

- $K^6 = E$, where the tensor $K = (K_{AB})$ is considered as a matrix,
- the eigenvalues of the tensor $K = (K_{AB})$ are $1, 1, 1, q, -q$.

Definition 1. *A third-order hypermatrix $T = (T_{ijk})$ is said to be I_2-regular if the second SO(3)-invariant $I_2(T) = T_{ijk}T_{ikj}$ is non-zero.*

Obviously if T is a third-order traceless q-cyclic hypermatrix and z^A are the coordinates of this hypermatrix in the orthonormal basis (18) then T is I_2-regular if $K(z,z) \neq 0$.

5. Right Biunits of Ternary Algebra of Hypermatrices

Now our aim is to show that each I_2-regular traceless q-cyclic hypermatrix U can be used to construct a right biunit for the ternary product $T \diamond V \diamond W$. To this end, we will derive a more general formula for the ternary product $T \diamond T(u) \diamond T(v)$, where

T is an arbitrary third-order hypermatrix $T \in \mathfrak{T}^3$ and $T(u), T(v)$ are traceless q-cyclic hypermatrices with coordinates $(u^A), (v^B)$ respectively. It is convenient to split the calculation of the ternary product into two stages. Let us remind that

$$(T \diamond U \diamond V)_{ijk} = T_{ijp}\, U_{rsp}\, V_{srk}.$$

It is easy to see that U, V part of this product can be considered as the trace of the product of two square matrices $U^{(3)}_p = (U_{msp}), V^{(3)}_k = (V_{snk})$. Thus it is useful to introduce an auxiliary third-order square matrix H as follows

$$H = (H_{pk}), \quad H_{pk} = \mathrm{Tr}\,(U^{(3)}_p\, V^{(3)}_k).$$

Now the calculation of the entries of the ternary product $T \diamond U \diamond V$ can be performed by sequentially taking the rows of three square matrices in (10) and multiplying them on the right by the auxiliary matrix H. The three entries obtained in this way will be the row of the ternary product, that is,

$$(T \diamond U \diamond V)_{ijk} = T_{ijp}\, H_{pk}. \tag{25}$$

First of all, we write down the square matrices $U^{(3)}_p, V^{(3)}_k$. As we have already mentioned, for this we must take three columns of square matrices in (10) (the first, second and third columns, respectively), form a matrix from them and transpose it. We obtain

$$U^{(3)}_1 = \frac{1}{\sqrt{6}} \begin{pmatrix} 0 & -q\,u^2 & q\,u^3 \\ -u^2 & \bar{q}\,u^1 & \sqrt{2}\,\bar{q}\,u^4 \\ u^3 & \sqrt{2}\,u^5 & -\bar{q}\,u^1 \end{pmatrix}, \quad V^{(3)}_1 = \frac{1}{\sqrt{6}} \begin{pmatrix} 0 & -q\,v^2 & q\,v^3 \\ -v^2 & \bar{q}\,v^1 & \sqrt{2}\,\bar{q}\,v^4 \\ v^3 & \sqrt{2}\,v^5 & -\bar{q}\,v^1 \end{pmatrix},$$

$$U^{(3)}_2 = \frac{1}{\sqrt{6}} \begin{pmatrix} -\bar{q}\,u^2 & u^1 & \sqrt{2}\,q\,u^5 \\ q\,u^1 & 0 & -q\,u^3 \\ \sqrt{2}\,q\,u^4 & -u^3 & \bar{q}\,u^2 \end{pmatrix}, \quad V^{(3)}_2 = \frac{1}{\sqrt{6}} \begin{pmatrix} -\bar{q}\,v^2 & v^1 & \sqrt{2}\,q\,v^5 \\ q\,v^1 & 0 & -q\,v^3 \\ \sqrt{2}\,q\,v^4 & -v^3 & \bar{q}\,v^2 \end{pmatrix},$$

$$U^{(3)}_3 = \frac{1}{\sqrt{6}} \begin{pmatrix} \bar{q}\,u^3 & \sqrt{2}\,u^4 & -u^1 \\ \sqrt{2}\,\bar{q}\,u^5 & -\bar{q}\,u^3 & u^2 \\ -q\,u^1 & q\,u^2 & 0 \end{pmatrix}, \quad V^{(3)}_3 = \frac{1}{\sqrt{6}} \begin{pmatrix} \bar{q}\,v^3 & \sqrt{2}\,v^4 & -v^1 \\ \sqrt{2}\,\bar{q}\,v^5 & -\bar{q}\,v^3 & v^2 \\ -q\,v^1 & q\,v^2 & 0 \end{pmatrix}.$$

Then taking the trace of the products of these matrices we find the entries of the auxiliary matrix H. The diagonal entries of H are all equal and multiple of the invariant bilinear form $K(u, v)$, i.e.

$$H_{pp} = \frac{q}{3}(u^1v^1 + u^2v^2 + u^3v^3 + q\,u^4v^5 + q\,u^5v^4) = \frac{q}{3}\,K(u,v), \;\; p = 1,2,3. \tag{26}$$

Non-diagonal entries $H_{pk}, p \neq k$ can be written as follows

$$H_{12} = \frac{q}{6} \begin{vmatrix} u^2 & u^1 \\ v^2 & v^1 \end{vmatrix} + \frac{\bar{q}}{3\sqrt{2}} \begin{vmatrix} u^3 & u^4 \\ v^3 & v^4 \end{vmatrix} + \frac{q}{3\sqrt{2}} \begin{vmatrix} u^3 & u^5 \\ v^3 & v^5 \end{vmatrix},$$

$$H_{23} = \frac{q}{6} \begin{vmatrix} u^3 & u^2 \\ v^3 & v^2 \end{vmatrix} + \frac{q}{3\sqrt{2}} \begin{vmatrix} u^1 & u^4 \\ v^1 & v^4 \end{vmatrix} + \frac{\bar{q}}{3\sqrt{2}} \begin{vmatrix} u^1 & u^5 \\ v^1 & v^5 \end{vmatrix},$$

$$H_{31} = \frac{q}{6} \begin{vmatrix} u^1 & u^3 \\ v^1 & v^3 \end{vmatrix} + \frac{1}{3\sqrt{2}} \begin{vmatrix} u^2 & u^4 \\ v^2 & v^4 \end{vmatrix} + \frac{1}{3\sqrt{2}} \begin{vmatrix} u^2 & u^5 \\ v^2 & v^5 \end{vmatrix},$$

and the entries H_{21}, H_{32}, H_{13} are obtained by simultaneously rearranging the columns in each of the second order determinants in H_{12}, H_{23}, H_{31} respectively. Hence $H_{pk} = -H_{kp}$. The non-diagonal elements of the matrix H_{pk} can be written in compact form if we introduce the following third-order hypermatrix τ, which can be considered a q-analog of the Levi-Civita symbol. We define τ as follows

$$\tau_{ijk} = \begin{cases} 0, \text{ if there are at least two equal subscripts among } i, j, k \\ \tau_{ijk} = q\,\tau_{jki} \text{ for any even permutation of integers } 1, 2, 3 \\ \tau_{ijk} = \bar{q}\,\tau_{jki} \text{ for any odd permutation of integers } 1, 2, 3 \\ \tau_{312} = 1, \ \tau_{132} = -1. \end{cases} \tag{27}$$

From this definition we can easily find all the entries of the hypermatrix τ

$$\begin{aligned} \tau_{123} &= \bar{q}, \ \tau_{231} = q, \ \tau_{312} = 1, \\ \tau_{213} &= -\bar{q}, \ \tau_{321} = -q, \ \tau_{132} = -1. \end{aligned}$$

Thus, the structure of hypermatrix τ with respect to permutations of subscripts is in some way a mixture of a q-cyclic structure and a $\bar{q}$-cyclic structure, that is, by cyclically permuting an even permutation of integers 1,2,3, the entries of the hypermatrix τ are transformed according to the q-cyclic law, and in the case of an odd permutation we have a $\bar{q}$-cyclic law. In addition, the hypermatrix τ is skew-symmetric in the first two subscripts, that is, $\tau_{ijk} = -\tau_{jik}$. Moreover, if i, j, k is an even permutation of $1, 2, 3$ then $\tau_{kji} = -\bar{q}\tau_{ijk}, \tau_{ikj} = -q\,\tau_{ijk}$. Now making use of the hypermatrix τ we can write the non-diagonal entries of the matrix H as follows

$$H_{pk} = \frac{q}{6} \begin{vmatrix} u^k & u^p \\ v^k & v^p \end{vmatrix} + \frac{1}{3\sqrt{2}} \tau_{pkr} \begin{vmatrix} u^r & u^4 \\ v^r & v^4 \end{vmatrix} + \frac{1}{3\sqrt{2}} \bar{\tau}_{pkr} \begin{vmatrix} u^r & u^5 \\ v^r & v^5 \end{vmatrix} \tag{28}$$

Thus, these calculations show that the diagonal elements of matrix H form its symmetric part, which we will denote by $H_{\mathtt{symm}}$, and the non-diagonal elements form its skew-symmetric part, which we will denote by $H_{\mathtt{skew}}$. Then $H = H_{\mathtt{symm}} + H_{\mathtt{skew}}$. It is easy to see that the symmetric part of matrix H is determined by the bilinear form $K(u, v)$, while the skew-symmetric part is determined by the bivector constructed by means of two vectors u, v of the 5-dimensional complex space. It should be noted that the bilinear form $K(u, v)$ is non-degenerate, and therefore it determines the structure of the Clifford algebra on the 5-dimensional space of traceless q-cyclic hypermatrices.

Theorem 2. *Let U be a third-order traceless q-cyclic I_2-regular hypermatrix. Then the hypermatrix*

$$\hat{U} = \sqrt{\frac{3}{qI_2(U)}}\, U,$$

is a right biunit of the ternary algebra $(\mathfrak{T}^3, \diamond)$; that is, for any third-order hypermatrix T we have $T \diamond \hat{U} \diamond \hat{U} = T$.

Proof. Let T be a third-order hypermatrix, U, V be two third-order traceless q-cyclic hypermatrices whose coordinates in the basis (18) are u^A, v^B respectively. Then

$$(T \diamond U \diamond V)_{ijk} = T_{ijp}\, H_{pk},$$

where

$$H_{pk} = \frac{q}{6}\begin{vmatrix} u^k & u^p \\ v^k & v^p \end{vmatrix} + \frac{1}{3\sqrt{2}}\tau_{pkr}\begin{vmatrix} u^r & u^4 \\ v^r & v^4 \end{vmatrix} + \frac{1}{3\sqrt{2}}\bar{\tau}_{pkr}\begin{vmatrix} u^r & u^5 \\ v^r & v^5 \end{vmatrix}$$

Now if we assume $U = V$, that is $u^A = v^A$, then the skew-symmetric part of the auxiliary matrix H vanishes and we are left with the symmetric part

$$H_{pk} = \frac{q}{3}\, K(u, v)\, \delta_{pk}.$$

Hence

$$(T \diamond U \diamond U)_{ijk} = T_{ijp}\, \frac{q}{3}\, K(u, u)\, \delta_{pk} = \frac{q}{3}\, K(u, u)\, T_{ijk},$$

or

$$T \diamond U \diamond U = \frac{q I_2(U)}{3}\, T,$$

where we used $K(u, u) = I_2(U)$ and $I_2(U)$ is the value of the second invariant I_2 calculated in the case of a hypermatrix U. Since U is I_2-regular hypermatrix, that is, $I_2(U) \neq 0$, we can consider the hypermatrix

$$\hat{U} = \sqrt{\frac{3}{q I_2(U)}}\; U,$$

which clearly satisfies

$$T \diamond \hat{U} \diamond \hat{U} = T \diamond \Big(\sqrt{\frac{3}{q I_2(U)}}\; U\Big) \diamond \Big(\sqrt{\frac{3}{q I_2(U)}}\; U\Big) = \frac{3}{q I_2(U)}\; T \diamond U \diamond U = T$$

□

References

[1] Abramov, V., Kerner, R. and Le Roy, B. Hypersymmetry: A Z_3-generalization of supersymmetry, *J. Math. Phys.*, **38**, no. 3 (1997), 1650–1669.

[2] Abramov, V., Kerner, R., Shitov, S. and Liivapuu, O., Algebras with ternary law of composition and their realization by cubic matrices, *Jornal of Generalized Lie Theory and Applications*, **3**, no. 2 (2009), 77–94.

[3] Abramov, V., Liivapuu, O., SO(3)-Irreducible Geometry in Complex Dimension Five and ernary Generalization of Pauli Exclusion Principle, *Universe*, **10**, no. 2 (2024), https://doi.org/10.3390/universe10010002.

[4] Ahmad, F. Invariants of a Cartesian tensor of rank 3. *Arch. Mech.*, **63**, no. 10 (2011), 383–392.

[5] Bagger, J. and Lambert, N., Modeling multiple M2's, *Phys. Rev.* **D75** (2007), 045020, arXiv:hep-th/0611108.

[6] Bagger, J. and Lambert, N., Gauge symmetry and supersymmetry of multiple M2-branes, *Phys. Rev.* **D77** (2008), 065008, arXiv:0711.0955 [hep-th].

[7] Bars, I. and Günaydin, M., Dynamical theory of subconstituents based on ternary algebras, *Phys. Rev. D* **22**, no. 6 (1980), 1403 – 1413.

[8] Cherkis, S. and Sämann, C., Multiple M2-branes and generalized 3-Lie algebras, em Phys. Rev. D, **78** (2008), 066019.

[9] Filippov, V. T., n-Lie algebras, *Siberian Math. J.*, **26** (1985), 879–891.

[10] Gelfand, I. M., Minlos, R. A., Shapiro, Z. Ya. *Representations of the Rotation and Lorentz Groups and Their Applications*. Dover Publications, Ins. Mineola, New York, 2018.

[11] Kerner, R., Graduation Z_3 et la racine cubique de l'opérateur de Dirac, *C. R. Acad. Sci. Paris*, **312** (1991), 191–196.

[12] Kerner, R., Ternary Generalization of Pauli's Principle and the Z_6-Graded Algebras, *Phys. At. Nucl.*, **80** (2017), 522–534.

[13] Nambu, Y. Generalized Hamiltonian mechanics, *Phys. Rev. D*, **7** (1973), 2405–2412.

[14] Sokolov, N. P. *Three-dimensional Matrices and Their Applications*, Moscow, 1960 (in Russian).

[15] Takhtajan, L., On foundation of generalized Nambu mechanics. *Commun. Math. Phys.*, **160** (1994), 295–315.

[16] Wagner, V. V., A ternary algebraic structures in the theory of coordinate structures, *Doklady Akademii nauk SSSR*, **81**, no. 6 (1951), 981–984.

[17] Wagner, V. V., The theory of generalized heaps and generalized groups, *Matematicheskii Sbornik*, **74**, no. 3 (1953), 545 – 632.

[18] Zapata-Carratalá, C., Arsiwalla, X. D., Beynon, T., Heaps of Fish: arrays, generalized associativity and heapoids, arXiv:2205.05456[math.RA] (to appear in *Theoretical Computer Science*).

Chapter 2

Negative-Energy and Tachyonic Solutions in Relativistic Equations*

Valeriy V. Dvoeglazov †
UAF, Universidad de Zacatecas, Zacatecas, Zac., México

Abstract

It is well known that the relativistic equations have acausal solutions, which have generally been ignored. This is particularly true for higher spins. We consider spin 1/2 and spin 1 in this talk. We analyze corresponding propagators which may indicate if a theory is local or non-local. Negative-energy and tachyonic solutions are also considered. The conclusions are paradoxical in both spins.

The algebraic characteristic equations are $Det(\hat{p}-m)=0$ and $Det(\hat{p}+m)=0$, $\hat{p}=p^{\mu}\gamma_{\mu}$ for $u-$ and $v-$ 4-spinors of the spin-1/2. They have solutions with $p_0=\pm E_p=\pm\sqrt{\mathbf{p}^2+m^2}$. The recent problems of superluminal neutrinos, negative-mass squared neutrinos, various schemes of oscillations including sterile neutrinos, require attention. The problem of lepton mass splitting (e,μ,τ) has a long history. This suggests that something was missing in the foundations of relativistic quantum theories. Modifications appear to be necessary in the Dirac sea concept, and in the even more sophisticated Stueckelberg concept of backward propagation in time. The Dirac sea concept is intrinsically related to the Pauli principle. However, the Pauli principle is intrinsically related to the Fermi statistics

---

*Presented at the Lomsadze Conference, Dec. 17-19, 2024, Uzhgorod, Ukraine. Online.
†Corresponding Author's Email: valeri@fisica.uaz.edu.mx

In: Mathematical Problems in Relativity, Gravitation, and Cosmology
Editor: Valeriy Dvoeglazov
ISBN: 979-8-89530-622-2

and the anticommutation relations of fermions. Recently, the concept of the *bi-orthonormality* has been proposed; the (anti) commutation relations and statistics are assumed to be different for *neutral* particles. We propose relevant modifications in the basics of relativistic quantum theory below. Next, Sakharov in 1967, Ref. [1], introduced the idea of two universes with opposite arrows of time, born from the same initial singularity (i.e. Big Bang). Next, Next, Debergh et al. constructed (within the framework of the present-day quantum field theory) negative-energy fields for spin-1/2 fermions, Ref. [2]. Currently, the predominating consensus is the existence of dark matter (DM) and dark energy (DE) paradigms. Possible particle candidates have been proposed for the DM, but to date, the search for these candidates has not been successful. "There is growing favor with the idea that new ideas need to be considered until an answer is found." "A paradigm shift that allows serious consideration of negative mass is a real possibility." However, see Chubykalo an Vlayev, Ref. [3] on the relation of inertial and gravitational masses.

The general scheme for construction of the field operator has been presented in [4]. In the case of the $(1/2,0)\oplus(0,1/2)$ representation we have:

$$
\begin{aligned}
\Psi(x) &= \frac{1}{(2\pi)^3}\int d^4p\,\delta(p^2-m^2)e^{-ip\cdot x}\Psi(p) = \\
&= \frac{1}{(2\pi)^3}\sum_h\int d^4p\,\delta(p_0^2-E_p^2)e^{-ip\cdot x}u_h(p_0,\mathbf{p})a_h(p_0,\mathbf{p}) = \qquad (1)\\
&= \frac{1}{(2\pi)^3}\int\frac{d^4p}{2E_p}[\delta(p_0-E_p)+\delta(p_0+E_p)]\\
&\quad [\theta(p_0)+\theta(-p_0)]e^{-ip\cdot x}\sum_h u_h(p)a_h(p) = \\
&= \frac{1}{(2\pi)^3}\sum_h\int\frac{d^4p}{2E_p}[\delta(p_0-E_p)+\delta(p_0+E_p)]\\
&\quad \left[\theta(p_0)u_h(p)a_h(p)e^{-ip\cdot x}+\theta(p_0)u_h(-p)a_h(-p)e^{+ip\cdot x}\right]\\
&= \frac{1}{(2\pi)^3}\sum_h\int\frac{d^3\mathbf{p}}{2E_p}\theta(p_0)\\
&\quad \left[u_h(p)a_h(p)|_{p_0=E_p}e^{-i(E_pt-\mathbf{p}\cdot\mathbf{x})}+u_h(-p)a_h(-p)|_{p_0=E_p}e^{+i(E_pt-\mathbf{p}\cdot\mathbf{x})}\right],
\end{aligned}
$$

where $a_h, b_h^\dagger$ are the annihilation/creation operators, and in textbook cases

$$
u_h(\mathbf{p}) = \begin{pmatrix}\exp(+\boldsymbol{\sigma}\cdot\boldsymbol{\varphi}/2)\phi_R^h(\mathbf{0})\\ \exp(-\boldsymbol{\sigma}\cdot\boldsymbol{\varphi}/2)\phi_L^h(\mathbf{0})\end{pmatrix} \qquad (2)
$$

$cosh(\varphi)=E_p/m$, $sinh(\varphi)=|\mathbf{p}|/m$. During the calculations above we had to represent $1=\theta(p_0)+\theta(-p_0)$ in order to get positive- and negative-frequency parts [5]. Moreover, in these calculations we did not assume which equation this field operator (namely, the $u-$ spinor) satisfies, with negative- or positive- mass? Next, since bispinors are, in general, complex-valued, we can even use a different basis, such as $u_\alpha = column(i\,0\,0\,0)$ etc., instead of the customary one. In the Dirac case we should assume the following relation in the field operator:

$$\sum_h v_h(p) b_h^\dagger(p) = \sum_h u_h(-p) a_h(-p)\,. \tag{3}$$

We need $\Lambda_{\mu\lambda}(p) = \bar{v}_\mu(p) u_\lambda(-p)$. By direct calculations, we find

$$-m b_\mu^\dagger(p) = \sum_\lambda \Lambda_{\mu\lambda}(p) a_\lambda(-p)\,. \tag{4}$$

Hence, $\Lambda_{\mu\lambda} = -im(\boldsymbol{\sigma}\cdot\mathbf{n})_{\mu\lambda}$, $\mathbf{n} = \mathbf{p}/|\mathbf{p}|$. In the $(1,0)\oplus(0,1)$ representation a similar procedure leads to a different situation:

$$a_\mu(p) = [1 - 2(\mathbf{S}\cdot\mathbf{n})^2]_{\mu\lambda} a_\lambda(-p)\,. \tag{5}$$

This signifies that in order to construct the Sankaranarayanan-Good field operator [6] to satisfy $[\gamma_{\mu\nu}\partial_\mu\partial_\nu - \frac{(i\partial/\partial t)}{E} m^2]\Psi(x) = 0$, we need additional postulates.

We have, in fact, $u_h(E_p, \mathbf{p})$ and $u_h(-E_p, \mathbf{p})$ originally, which satisfy the equations: $[E_p(\pm\gamma^0) - \boldsymbol{\gamma}\cdot\mathbf{p} - m]\, u_h(\pm E_p, \mathbf{p}) = 0$. Due to the properties $U^\dagger\gamma^0 U = -\gamma^0$, $U^\dagger\gamma^i U = +\gamma^i$ with the unitary matrix $U = \gamma^0\gamma^5$ we have in the negative-energy case:

$$\left[E_p\gamma^0 - \boldsymbol{\gamma}\cdot\mathbf{p} - m\right] U^\dagger u_h(-E_p, \mathbf{p}) = 0\,. \tag{6}$$

The explicit forms $\gamma^5\gamma^0 u(-E_p, \mathbf{p})$ are different from the textbook "positive-energy" Dirac spinors. After the space inversion operation, we have ($R = (\mathbf{x} \to -\mathbf{x}, \mathbf{p} \to -\mathbf{p})$)

$$PR\tilde{u}(p) = PR\gamma^5\gamma^0 u_\uparrow(-E_p, \mathbf{p}) = -\tilde{u}(p)\,, \tag{7}$$

$$PR\tilde{\tilde{u}}(p) = PR\gamma^5\gamma^0 u_\downarrow(-E_p, \mathbf{p}) = -\tilde{\tilde{u}}(p)\,. \tag{8}$$

Similar formulations have been presented in Refs. [7], and [8]. The group-theoretical basis for such doubling has been given in the papers by Gelfand, Tsetlin and Sokolik [9], who first presented the theory in the 2-dimensional representation of the inversion group in 1956 (later called "the Bargmann-Wightman-Wigner-type quantum field theory" in 1993). Barut and Ziino [8] proposed yet another model. They considered the γ^5 operator to be the operator of charge conjugation. Thus, the charge-conjugated Dirac equation has a different sign compared to the ordinary formulation:

$$[i\gamma^\mu\partial_\mu + m]\Psi^c_{BZ} = 0\,, \tag{9}$$

and the charge conjugation so defined applies to the whole system, fermion+electromagnetic field, $e \to -e$ in the covariant derivative. The superpositions of the Ψ_{BZ} and Ψ^c_{BZ} also give us the "doubled Dirac equation", the equations for $\lambda-$ and $\rho-$ self/anti-self charge conjugate spinors. The concept of doubling the Fock space has been developed in the Ziino program (cf. Refs. [9, 10]) within the framework of the quantum field theory. In the BZ case the charge conjugate states are simultaneously the eigenstates of the chirality. Here, the relevant paper is Ref. [11]. It is straightforward to merge $u(\mathbf{p})$ and $v(\mathbf{p})$ spinors in one doublet of "positive energy" and $v(\mathbf{p})$ and $u(\mathbf{p})$ spinors, in another doublet of "negative energy" , as Markov and Fabbri did. However, the point of my paper is that both $u(p_0, \mathbf{p})$ and $v(p_0, \mathbf{p})$ contains contributions to both positive- and negative- energies, cf. Ref. [12].

We study the problem of construction of causal propagators in spin $S = 1/2$ and higher-spin theories. The hypothesis is: in order to construct analogues of the Feynman-Dyson propagator we actually need four field operators connected by the dual and parity transformation. We use the standard methods of quantum field theory. Thus, the number of components in the *causal* propagators is enlarged accordingly. The conclusion under discussion is that if we did not expand the number of components in the fields (in the propagator) we would not be able to obtain the causal propagator.

According to the Feynman-Dyson-Stueckelberg conception, the $S = 1/2$ causal propagator S_F has to be constructed on using the formula (*e.g.*, Ref. [13, p.91])

$$\begin{aligned} S_F(x_2, x_1) = \sum_\sigma \int \frac{d^3p}{(2\pi)^3} \frac{m}{E_p} \\ \times \left[\theta(t_2 - t_1)\, a\; u^\sigma(p) \overline{u}^\sigma(p) e^{-ip\cdot x} \right. \\ \left. + \theta(t_1 - t_2)\, b\; v^\sigma(p) \overline{v}^\sigma(p) e^{ip\cdot x}\right], \end{aligned} \tag{10}$$

where $x = x_2 - x_1$. In the spin $S = 1/2$ Dirac theory, it results in

$$S_F(x) = \int \frac{d^4p}{(2\pi)^4} e^{-ip\cdot x} \frac{\hat{p} + m}{p^2 - m^2 + i\epsilon}, \tag{11}$$

provided that the constants a and b are determined by imposing

$$(i\hat{\partial}_2 - m) S_F(x_2, x_1) = \delta^{(4)}(x_2 - x_1), \tag{12}$$

namely, $a = -b = 1/i$; $\partial_2 = \partial/\partial x_2$, ϵ defines the rules of work near the poles.

However, attempts to construct the causal covariant propagator in this way failed in the framework of the Weinberg theory, Ref. [14], which is a generalization of

Dirac's ideas to higher spins. The propagator proposed in Ref. [15] is the causal propagator. However, the old problem remains: the Feynman-Dyson propagator is not the Green function of the Weinberg equation. As mentioned, the covariant propagator proposed by Weinberg propagates kinematically spurious solutions [15]. We construct the propagator in the framework of the model given in Ref. [10]. The concept of the Weinberg field *doubles* has been proposed there. For the functions $\,^{(1)}_{1}$ and $\,^{(1)}_{2}$, connected with the former by the dual (chiral, $\gamma_5 = diag(1_{3\times 3}), -1_{3\times 3}))$ transformation, the equations are[1]

$$(\gamma_{\mu\nu}p_\mu p_\nu + m^2)\psi_1^{(1)} = 0\,, \tag{13}$$

$$(\gamma_{\mu\nu}p_\mu p_\nu - m^2)\psi_2^{(1)} = 0\,, \tag{14}$$

with $\mu, \nu = 1, 2, 3, 4$. For the field functions connected with $\psi_1^{(1)}$ and $\psi_2^{(1)}$ by the $\gamma_5\gamma_{44}$ transformations the set of equations is written:

$$\left[\widetilde{\gamma}_{\mu\nu}p_\mu p_\nu - m^2\right] \,^{(2)}_{1} = 0\,, \tag{15}$$

$$\left[\widetilde{\gamma}_{\mu\nu}p_\mu p_\nu + m^2\right] \,^{(2)}_{2} = 0\,, \tag{16}$$

where $\widetilde{\gamma}_{\mu\nu} = \gamma_{44}\gamma_{\mu\nu}\gamma_{44}$ is connected with the $S = 1$ Barut-Muzinich-Williams $\gamma_{\mu\nu}$ matrices [17]. In the cited paper I have used the plane-wave expansion. Thus, $u_1^{(2)}(\mathbf{p}) = \gamma_5\gamma_{44}u_1^{(1)}(\mathbf{p})$, $\overline{u}_1^{(2)} = \overline{u}_1^{(1)}\gamma_5\gamma_{44}$, $u_2^{(2)}(\mathbf{p}) = \gamma_5\gamma_{44}\gamma_5 u_1^{(1)}(\mathbf{p})$ and $\overline{u}_2^{(2)}(\mathbf{p}) = -\overline{u}_1^{(1)}\gamma_{44}$. Now we check whether the sum of the four equations

$$\begin{aligned}
&\left[\gamma_{\mu\nu}\partial_\mu\partial_\nu - m^2\right] * \\
&* \int \frac{d^3p}{(2\pi)^3 2E_p}\left[\theta(t_2 - t_1)\, a\;\; u_1^{\sigma\,(1)}(p)\overline{u}_1^{\sigma\,(1)}(p)e^{ip\cdot x} + \theta(t_1 - t_2)\, b\;\; v_1^{\sigma\,(1)}(p)\overline{v}_1^{\sigma\,(1)}(p)e^{-ip\cdot x}\right] + \\
&\left[\gamma_{\mu\nu}\partial_\mu\partial_\nu + m^2\right] * \\
&* \int \frac{d^3p}{(2\pi)^3 2E_p}\left[\theta(t_2 - t_1)\, a\;\; u_2^{\sigma\,(1)}(p)\overline{u}_2^{\sigma\,(1)}(p)e^{ip\cdot x} + \theta(t_1 - t_2)\, b\;\; v_2^{\sigma\,(1)}(p)\overline{v}_2^{\sigma\,(1)}(p)e^{-ip\cdot x}\right] + \\
&\left[\widetilde{\gamma}_{\mu\nu}\partial_\mu\partial_\nu + m^2\right] * \\
&* \int \frac{d^3p}{(2\pi)^3 2E_p}\left[\theta(t_2 - t_1)\, a\;\; u_1^{\sigma\,(2)}(p)\overline{u}_1^{\sigma\,(2)}(p)e^{ip\cdot x} + \theta(t_1 - t_2)\, b\;\; v_1^{\sigma\,(2)}(p)\overline{v}_1^{\sigma\,(2)}(p)e^{-ip\cdot x}\right] + \\
&\left[\widetilde{\gamma}_{\mu\nu}\partial_\mu\partial_\nu - m^2\right] * \\
&* \int \frac{d^3p}{(2\pi)^3 2E_p}\left[\theta(t_2 - t_1)\, a\;\; u_2^{\sigma\,(2)}(p)\overline{u}_2^{\sigma\,(2)}(p)e^{ip\cdot x} + \theta(t_1 - t_2)\, b v_2^{\sigma\,(2)}(p)\overline{v}_2^{\sigma\,(2)}(p)e^{-i\cdot px}\right] \\
&= \;\; \delta^{(4)}(x_2 - x_1)
\end{aligned} \tag{17}$$

[1]I have to use the Euclidean metrics here in order a reader to be able to compare the formalism with the classical cited works.

can be satisfied by a definite choice of a and b. Simple calculations give

$$\begin{aligned}
&\partial_\mu \partial_\nu \left[a\,\theta(t_2 - t_1)\, e^{ip(x_2-x_1)} + b\,\theta(t_1 - t_2)\, e^{-ip(x_2-x_1)} \right] = \\
&-\left[a\, p_\mu p_\nu \theta(t_2 - t_1) \exp\left[ip(x_2 - x_1) \right] + b\, p_\mu p_\nu \theta(t_1 - t_2) \exp\left[-ip(x_2 - x_1) \right] \right] \\
&+a\left[-\delta_{\mu 4}\delta_{\nu 4}\delta\,'(t_2 - t_1) + i(p_\mu \delta_{\nu 4} + p_\nu \delta_{\mu 4})\delta(t_2 - t_1) \right] \\
&\exp\left[i\mathbf{p}\cdot(\mathbf{x}_2 - \mathbf{x}_1) \right] + b\, \left[\delta_{\mu 4}\delta_{\nu 4}\delta\,'(t_2 - t_1) + \right. \\
&\left. i(p_\mu \delta_{\nu 4} + p_\nu \delta_{\mu 4})\delta(t_2 - t_1) \right] \exp\left[-i\mathbf{p}(\mathbf{x}_2 - \mathbf{x}_1) \right]\; ;
\end{aligned} \tag{18}$$

We conclude as follows: the generalization of the notion of causal propagators is admitted by the use of the Wick-like formula for the time-ordered particle operators provided that $a = b = 1/4im^2$. It is necessary to consider all four equations, Eqs. (13)-(16). Obviously, this is related to the 12-component formalism, which I presented in Ref. [10].

The $S = 1$ analogues of the formula (11) for the Weinberg propagators follow immediately. In the Euclidean metrics they are:

$$S_F^{(1)}(p) \sim -\frac{1}{i(2\pi)^4(p^2+m^2-i\epsilon)} \left[\gamma_{\mu\nu} p_\mu p_\nu - m^2 \right], \tag{19}$$

$$S_F^{(2)}(p) \sim -\frac{1}{i(2\pi)^4(p^2+m^2-i\epsilon)} \left[\gamma_{\mu\nu} p_\mu p_\nu + m^2 \right], \tag{20}$$

$$S_F^{(3)}(p) \sim -\frac{1}{i(2\pi)^4(p^2+m^2-i\epsilon)} \left[\widetilde{\gamma}_{\mu\nu} p_\mu p_\nu + m^2 \right], \tag{21}$$

$$S_F^{(4)}(p) \sim -\frac{1}{i(2\pi)^4(p^2+m^2-i\epsilon)} \left[\widetilde{\gamma}_{\mu\nu} p_\mu p_\nu - m^2 \right]. \tag{22}$$

Meanwhile, I propose to use the 8-component (or 16-component) spin-1/2 formalism in similarity with the 12-component formalism of this discussion. If we calculate

$$\begin{aligned}
S_F^{(+,-)}(x_2, x_1) = &\int \frac{d^3p}{(2\pi)^3} \frac{m}{E_p} \left[\theta(t_2 - t_1)\, a\; \Psi^\sigma_\pm(p) \overline{\Psi}^\sigma_\pm(p) e^{-ip\cdot x} \right. \\
&\left. + \theta(t_1 - t_2)\, b\; \Psi^\sigma_\mp(p) \overline{\Psi}^\sigma_\mp(p) e^{ip\cdot x} \right] \\
= &\int \frac{d^4p}{(2\pi)^4} e^{-ip\cdot x} \frac{(\hat{p} \pm m)}{p^2 - m^2 + i\epsilon},
\end{aligned} \tag{23}$$

(with Ψ doublets in the field operator) we readily come to the result that the corresponding Feynman-Dyson propagator gives the local theory in the sense:

$$\sum_{\pm} [i\Gamma_\mu \partial_2^\mu \mp m] S_F^{(+,-)}(x_2 - x_1) = \delta^{(4)}(x_2 - x_1), \tag{24}$$

even in the case of self/anti-self charge conjugate states. However, physics should choose only one correct formalism. It is not clear, why two correct mathematical formalisms lead to different physical results (e.g., Rodrigues, Jr. and Ahluwalia). First of all, we should check whether possible non-locality in the propagators has an influence on the physical observables such as the scattering amplitudes, the energy spectra and the decay widths. If not, we may find some unexpected symmetries in relativistic quantum mechanics/field theory. This is a task for future publications. However, it is already obvious that if we did not enlarge the number of components in the fields (in the propagator), we would not be able to obtain the formally causal propagators for higher spins and/or for neutral particles. Using these ideas, it is easy (but somewhat cumbersome) to construct propagators for other spins.

The dilemma of the (non)local propagators for the spin $S = 1$ has also been analyzed in Ref. [16] within the Duffin-Kemmer-Petiau (DKP) formalism or the Dirac-Kähler formalism [16]. However, the propagators given in Ref. [16] are actually those in the generalized Duffin-Kemmer-Petiau formalism. They are not in the Weinberg-Tucker-Hammer formalism. Moreover, the problem of the massless limit was not discussed in the DKP formalism, which is non-trivial (like that of the Proca formalism [20]).

We should use the set of Weinberg propagators obtained (19-22) in the perturbation calculus of scattering amplitudes. In Ref. [18] the amplitude for the interaction of two $2(2S+1)$ bosons has been obtained on the basis of the use of one field only, and it is obviously incomplete, see also Ref. [17]. But, it is interesting to note that the spin structure was proved there to be the same, whether we consider the two-Dirac-fermion interaction or the two-Weinberg $S=1$-boson interaction. However, the denominator differs slightly $(1/\vec{\Delta}^2 \rightarrow 1/2m(\Delta_0 - m))$ from the fermion-fermion case in the cited papers [18], where $\Delta_0, \vec{\Delta}$ is the momentum-transfer 4-vector in Lobachevsky space. More accurate considerations of the fermion-boson and boson-boson interactions in the framework of the Weinberg theory have been reported elsewhere, Ref. [19]. So, the conclusion is: one can construct analogs of the Feynman-Dyson propagators for the $2(2S+1)$ model and, hence also local theories, provided that the Weinberg states are quadrupled ($S = 1$ case), and the neutral particle states are doubled.

What is the physical sense of the mathematical formalism presented here? Why did we consider four field functions in the propagator for spin-1 in Ref. [10]? We have 4 solutions in the original Dirac equation for $u-$ and 4 solutions for $v = \gamma^5 u$ (recall that $p_0 = \pm E_p = \pm\sqrt{\mathbf{p}^2 + m^2}$). See, for example, Ref. [21]. In the $S = 1$ Weinberg equation [14] we have 12 solutions.[2] Apart from $p_0 = \pm E_p$ we have tachyonic solutions $p_0 = \pm E_p' = \pm\sqrt{\mathbf{p}^2 - m^2}$, *i. e.* $m \rightarrow im$. This is easily

[2]In Ref. [17] we have causal solutions only for the S=1 Tucker-Hammer equation.

checked by using the algebraic equations and solving them with respect to p_0:

$$Det[\gamma^{\mu}p_{\mu} \pm m] = 0\,, \tag{25}$$

and

$$Det[\gamma^{\mu\nu}p_{\mu}p_{\nu} \pm m^2] = 0\,. \tag{26}$$

In constructing the field operato, Ref. [21] we generally need $u(-p) = u(-p_0, -\mathbf{p}, m)$ which should be transformed to $v(p) = \gamma^5 u(p) = \gamma^5 u(+p_0, +\mathbf{p}, m)$. On the other hand, when we calculate the parity properties we need $\mathbf{p} \to -\mathbf{p}$. The $u(p_0, -\mathbf{p}, m)$ satisfies

$$[\tilde{\gamma}^{\mu\nu}p_{\mu}p_{\nu} + m^2]u(p_0, -\mathbf{p}, m) = 0\,. \tag{27}$$

The $u(-p_0, \mathbf{p}, m)$ “spinor" satisfies:

$$[\tilde{\gamma}^{\mu\nu}p_{\mu}p_{\nu} + m^2]u(-p_0, +\mathbf{p}, m) = 0\,, \tag{28}$$

that is the same as above. The tilde signifies $\tilde{\gamma}^{\mu\nu} = \gamma_{00}\gamma^{\mu\nu}\gamma_{00}$ that is analogoues to the $S = 1/2$ case $\tilde{\gamma}^{\mu} = \gamma_0\gamma^{\mu}\gamma_0$. The $u(-p_0, -\mathbf{p}, m)$ satisfies:

$$[\gamma^{\mu\nu}p_{\mu}p_{\nu} + m^2]u(-p_0, -\mathbf{p}, m) = 0\,. \tag{29}$$

This case is opposite to the spin-1/2 case where the spinor $u(-p_0, \mathbf{p}, m)$ satisfies

$$[\tilde{\gamma}^{\mu}p_{\mu} + m]u(-p_0, +\mathbf{p}, m) = 0\,, \tag{30}$$

and $u(p_0, -\mathbf{p}, m)$,

$$[\tilde{\gamma}^{\mu}p_{\mu} - m]u(p_0, -\mathbf{p}, m) = 0\,, \tag{31}$$

and

$$[\gamma^{\mu}p_{\mu} + m]u(-p_0, -\mathbf{p}, m) = 0\,. \tag{32}$$

In general we can use $u(-p_0, +\mathbf{p}, m)$ or $u(p_0, -\mathbf{p}, m)$ to construct the causal propagator in the spin-1/2 case. However, we do not need to use both because a) $u(-p_0, +\mathbf{p}, m)$ satisfies a similar equation to $u(+p_0, -\mathbf{p}, m)$ and b) we have an integration over $\mathbf{p}$. This integration is invariant with respect to $\mathbf{p} \to -\mathbf{p}$. So the physical result for the causal propagator does not change regardless of whether one or another Dirac equation is used. The situation is different for spin 1. The tachyonic solutions of the original Weinberg equation

$$[\gamma^{\mu\nu}p_{\mu}p_{\nu} + m^2]u(p_0, +\mathbf{p}, m) = 0 \tag{33}$$

are just the solutions of the equation with the opposite square of $m \to im$):

$$[\gamma^{\mu\nu}p_{\mu}p_{\nu} - m^2]u(p_0, +\mathbf{p}, im) = 0\,. \tag{34}$$

We cannot transform the propagator of the original equation (33) to that of (34) solely by a change of variables, as in the spin-1/2 case. The mass squared changes the sign, just as in the case of $v-$ "spinors". When we construct the propagator we have to take this solution into account, as well as the superposition $u(p, m)$ and $u(p, im)$, and corresponding equations. The conclusion is paradoxical: in order to construct the causal propagator for spin 1 we have to take acausal (tachyonic) solutions of homogeneous equations into account. It is not surprising that the propagator is not causal for the Tucker-Hammer equation because it does not contain the tachyonic solutions. This statement is probably valid for all higher spins.

Acknowledgments

I acknowledge discussions with Prof. N. Debergh. I am grateful to Zacatecas University for a professorship. I appreciate discussions with participants of several recent conferences.

References

[1] A. D. Sakharov, *JETP Lett.* **5**, 24 (1967).

[2] N. Debergh, J-P. Petit and G. D'Agostini, *J. Phys. Comm.* **2**, 115012 (2018).

[3] A. E. Chubykalo and S. J. Vlayev, *Eur. J. Phys.* **19**, 1 (1998).

[4] N. N. Bogoliubov and D. V. Shirkov, *Introduction to the Theory of Quantized Fields.* 2nd Edition. (Nauka, Moscow. 1973).

[5] V. V. Dvoeglazov, *J. Phys. Conf. Ser.* **284**, 012024 (2011), arXiv:1008.2242.

[6] A. Sankaranarayanan and R. H. Good, jr., *Nuovo Cim.* **36**, 1303 (1965).

[7] M. Markov, *ZhETF* **7**, 579 (1937); ibid. 603; *Nucl. Phys.* **55**, 130 (1964).

[8] A. Barut and G. Ziino, *Mod. Phys. Lett.* A**8**, 1011 (1993); G. Ziino, *Int. J. Mod. Phys.* A**11**, 2081 (1996).

[9] I. M. Gelfand and M. L. Tsetlin, *ZhETF* **31**, 1107 (1956); G. A. Sokolik, *ZhETF* **33**, 1515 (1957).

[10] V. V. Dvoeglazov, *Int. J. Theor. Phys.* **37**, 1915 (1998).

[11] L. Fabbri, *Int. J. Theor. Phys.* **53**, 1896 (2014), arXiv: 1210.1146.

[12] G.-J. Ni, in "*Relativity, Gravitation, Cosmology: New Developments*". (Nova Science Pubs., NY, USA, 2010), p. 253.

[13] C. Itzykson and J.-B. Zuber, *Quantum Field Theory.* (McGraw-Hill Book Co. New York, 1980).

[14] S. Weinberg, *Phys. Rev.* B**133**, 1318 (1964).

[15] D. V. Ahluwalia and D. J. Ernst, *Phys. Rev.* C**45**, 3010 (1992).

[16] S. I. Kruglov, *Int. J. Theor. Phys.* **41**, 653 (2002), arXiv:hep-th/0110251. S. I. Kruglov, in *"Einstein and Hilbert: Dark Matter"*. (Nova Science Pubs, NY, USA, 2012) p. 107, arXiv: 0912.3716.

[17] R. H. Tucker and C. L. Hammer, *Phys. Rev.* D**3**, 2448 (1971).

[18] V. V. Dvoeglazov, *Int. J. Theor. Phys.* **35**, 115 (1996).

[19] V. V. Dvoeglazov, *J. Phys. Conf. Ser.* **128**, 012002 (2008).

[20] A. V. Hradyskyi and Yu. P. Stepanovskiy, *Ukr. J. Phys.* **63**, 584 (2018).

[21] J. A. Cazares and V. V. Dvoeglazov, Rev. Mex. Fis. **69**, 050703 1-9 (2023).

Chapter 3

Approach for Modelling Quantum-Mechanical Collapse

A. Yu. Ignatiev *
Theoretical Physics Research Institute, Melbourne, Australia

Abstract

A long-standing quantum-mechanical puzzle is whether the collapse of the wave function is a real physical process or simply an epiphenomenon. This puzzle lies at the heart of the measurement problem. One way to choose between the alternatives is to assume that one or the other is correct and attempt to draw physical, observable consequences which could then be empirically verified or ruled out.

As a working hypothesis, we propose simple models of collapse as a real physical process for direct binary symmetric measurements made on one particle. This allows one to construct irreversible unstable Schrödinger equations capable of describing continuously the process of collapse induced by the interaction of the quantum system with the measuring device. Due to unknown initial conditions the collapse outcome remains unpredictable so no contradictions with quantum mechanics arise.

Our theoretical framework predicts a finite time-scale of the collapse and links with experiment. Sensitive probes of the collapse dynamics could be done using Bose-Einstein condensates, ultracold neutrons or ultrafast optics.

---

*Corresponding Author's Email: a.ignatiev@ritp.org

In: Mathematical Problems in Relativity, Gravitation, and Cosmology
Editor: Valeriy Dvoeglazov
ISBN: 979-8-89530-622-2

If confirmed, the formulation could be relevant to the transition from quantum fluctuations to classical inhomogeneities in early cosmology and to establishing the ultimate limits on the speed of quantum computation and information processing.

1. Introduction

Modern technology has turned in-depth studies of quantum mechanics into a fast-growing research area. The physical meaning and the evolution of quantum state are among the focal points in these studies [1, 2, 3].

The wave function collapse (also known as the reduction of the wave packet) is an irreversible, unpredictable change of quantum state which occurs during the measurement and which cannot be described by the reversible Schrödinger equation. The nature and physical mechanisms of collapse are among the most debated issues of quantum mechanics, and our understanding of this mysterious process remains rather limited.

Collapse has been discussed from numerous, often very different perspectives, starting from purely philosophical and ending with purely mathematical. Philosophical discussions focus on various interpretations of quantum mechanics, and point out that in some of them, collapse is not a physical process but rather a (trivial) mental process of updating the available information.

This approach is often illustrated with the example of picking balls from a box with one black and one white ball. Initially, the probability to pick the white ball is 1/2. But after the first picked ball turns out black, then the probability to pick up the white ball becomes 1. In this case, which is completely classical, even the term 'collapse' itself seems unnecessary. Indeed, before the advent of quantum mechanics, no-one was talking about collapses in the theory of probability.

Such a viewpoint emerged when the options for quantum experimenting were limited by the level of technology. Since then, completely new kinds of experiments became a reality, and it is now possible to consider the collapse problem afresh.

Besides, the quantum collapse without observers has become a vital issue in cosmology where a physical mechanism is needed for turning quantum fluctuations into classical, which would then evolve into galaxies.

As a consequence, the modern approach consists in translating philosophical questions into physical questions. In this vein, known as "experimental metaphysics", we do not postulate 'absence of collapse' or otherwise, but rather ask how we can find out if there is a collapse or not using ordinary physical methods and principles.

The problem of the collapse under measurement is intimately related (but not identical) to the problem of quantum-to-classical transition or the transition from possible to actual.

The theory of environmental decoherence has made impressive progress in this area over recent decades. Optimistically, a view was advocated by some authors that decoherence has solved the measurement problem and the concept of collapse has become superfluous.

However, this issue is far from being settled (see, e.g., [6] and references therein).

Conventional quantum mechanics models the transition from quantum to classical as a black box.

Decoherence theory has shone some light on that box, but do we know what is inside? What happens at the final, most intriguing stage of the measurement process? Only 15 percent of experts believe that the decoherence program has solved the measurement problem [7].

For this reason, it seems warranted to consider first the case of 'pure' collapse, i.e., neglect environmental decoherence altogether. This approach can be justified in the end by comparing the two characteristic time-scales: for collapse and for decoherence. If the former is much shorter than the latter, then decoherence indeed can be neglected. This case is also more interesting from a practical point of view, since in quantum computing decoherence needs to be reduced as much as possible.

2. Non-linear Effective Equations for Collapse

The problem of collapse during the measurement is essentially a many body problem (or even an infinitely many degrees of freedom problem). For this reason, exact equations governing the collapse process are hard to obtain, and even if obtained, would be difficult to use. In such cases, a common method is to eliminate the unwanted degrees of freedom by applying the phenomenological, effective theory approach. The essence of this approach is to 'integrate out' the unknown quantities. This means introducing new terms built out of the known quantities subject to certain symmetry requirements and multiplying them by unknown coefficients. This has become a standard practice of quantum field theory over the recent decades, but is less appreciated in the context of quantum mechanics.

When introducing new terms, our strategy is conservative, and we will only rely on the most general physical criteria: instability, irreversibility, and symmetry breaking. The first two features had already been advocated by Niels Bohr, while the third is among the most universal phenomena in contemporary physics.

We start with the familiar setting where collapse is supposed to occur: a quantum system with a wave function plus a measuring apparatus.

Following the spirit of the effective-theory, phenomenological type of approach, we set ourselves the following task: Take the Schrödinger equation and add new terms which are made entirely out of the wave function and unknown coefficients. The new equations should be able to describe both ordinary evolution and the measurement process (i.e., the collapse of the wave function).

This may sound like an extraordinary proposal which is bound to fail. Indeed, it seems that nothing could be so different as continuous, deterministic evolution and sudden, unpredictable collapse. Further, it seems unavoidable that some random, stochastic sources, or an additional physical phenomenon should necessarily be introduced to account for the collapse, as was done, for instance, in Refs.[8, 9, 10, 11, 12].

Nevertheless, we require from the outset that no new stochastic sources or gravitation should be invoked in explaining collapse[1].

The main difficulty in this search is the fact that quantum and classical systems essentially speak different languages and it is hard to make them talk to each other. The quantum language is based on operators while the classical one—on ordinary functions. To get around this obstacle, we can try to rewrite quantum mechanics in a more classical form, and thus to obtain a zeroth approximation suitable for our effective theory.

The most convenient for us is the approach known as the complex quantum Hamilton-Jacobi (CQHJ) formulation [13, 14, 15, 16, 17, 18] which has attracted growing attention recently. It is natural to expect that the new formulation could give us new insights into fundamental long-standing issues such as the mechanism of quantum collapse.

For brevity and simplicity, we focus first on the simplest yet typical case of the wave function collapse. This case is realised in a one-dimensional symmetric binary

[1] To avoid confusion, we emphasize that the Continuous Spontaneous Localization (CSL) theory of Ghirardi-Rimini-Weber and Pearle [8, 9, 10] and the gravity-induced collapse models, for instance, those of Penrose [11] and Diósi [12] are in principle different from our proposal because they address an essentially different problem: collapse during free evolution and not during a measurement. So it would be meaningless to ask which theory is better. More details are in Sec. V.

measurement of the momentum where only two outcomes are possible and the initial wave function is symmetric with respect to these outcomes.

It can be shown (see Refs.[13, 14, 15, 16, 17, 18, 19] and Appendix) that the Schrödinger equation in this case becomes

$$p_t = -\nabla H, \tag{1}$$

where

$$p(x,t) = \frac{\hbar}{i}\frac{\psi'}{\psi}, \quad H = V + \frac{p^2}{2m} + \frac{\hat{p}p}{2m}. \tag{2}$$

The parity-symmetric wave function before the collapse is the 1:1 superposition of two plane waves with opposite directions:

$$\psi = \frac{1}{\sqrt{2}}(e^{ikx} + e^{-ikx}). \tag{3}$$

We need to construct the phenomenological equation that would collapse this wave function into either $\psi_+ = e^{ikx}$ or $\psi_- = e^{-ikx}$. Following Bohr, we assume that the collapse is caused by the interaction of the quantum system with a semi-classical measuring device. Usually, the interaction is described by the Hamiltonian (or potential) that does not depend on the quantum state. However, such interactions cannot cause the collapse and should be extended to include state-dependent interactions. The phenomenological state-dependent potentials (or their analogues) are familiar in many areas of quantum theory (for instance, equations of Ginzburg-Landau, Gross-Pitaevsky, quantum chemistry etc.).

A state-dependent interaction is not a fundamental interaction, but an approximate one that is obtained, e.g., as a result of mean-field approximation which assumes a definite classical value for a mean field. So the proposed approach does not attempt to derive definite outcomes from a fundamental theory.

The state-dependent potential V_S should be added to the ordinary potential V in Eq. (22) to obtain:

$$p_t = -\nabla(H + V_S). \tag{4}$$

The gradient of the potential V_S is a more convenient quantity than the potential itself, and we introduce a special term for it. It is natural to refer to this gradient as some kind of generalized force; in the context of our problem, we will call it "the collapsing force":

$$F_c = -\nabla V_S. \tag{5}$$

The fact that the potential V_S should be state-dependent can be equivalently stated as the requirement that the collapsing force F_c should be p-dependent.

Due to the shortness of collapse, the interaction with the apparatus is expected to be dominant during the measurement process, so our Eq. (4) can be written in a simpler form

$$p_t = F_c. \tag{6}$$

To find the form of F_c, we invoke the general principles as formulated in the beginning:

(1) Symmetry;

(2) Irreversibility;

(3) Instability.

It is natural to start with the simplest kind of functions, such as low-degree polynomials (which can also be viewed as the first terms of a Taylor expansion):

$$F_c = c_0 + c_1 p + c_2 p^2 + c_3 p^3 + \dots \tag{7}$$

According to Eq. (6), the force must be a vector under parity symmetry ($p \to -p$) as expected. Therefore, the even-power terms in this expansion, c_0 and $c_2 p^2$, should be dropped as violating this symmetry.

The requirement that the measurement gives two outcomes, $p = +q$ or $p = -q$, translates into the condition that F_c must vanish at $p = \pm q$:

$$F_c(+q) = F_c(-q) = 0. \tag{8}$$

Therefore, we must have $c_1 = -c_3 q^2$, and introducing $g = -c_3$ we obtain

$$F_c = gp(q^2 - p^2). \tag{9}$$

Consequently, g is interpreted as the effective coupling constant controlling the strength of interaction between the system and apparatus.

In general, g could be a complex number, but in order to satisfy the condition of irreversibility we assume that it is real.

Altogether, our general equation (6) takes on a 'collapsible' form

$$p_t = gp(q^2 - p^2). \tag{10}$$

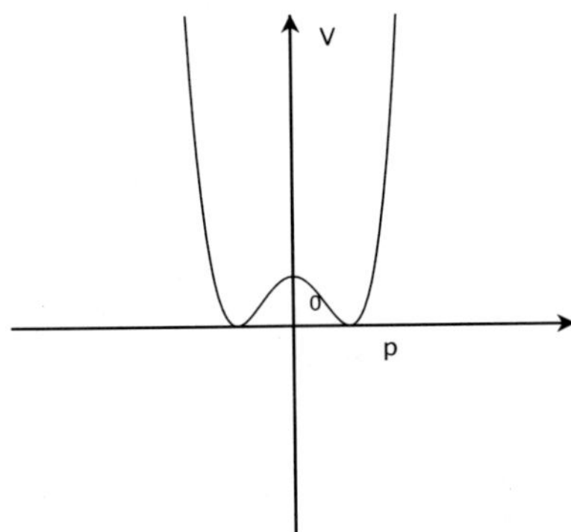

Figure 1. The schematic form of the 'collapsing potential' V(p). This 2-dimensional plot assumes that p and therefore V are real numbers. (In general, both are complex so a 4-dimensional plot would be required.)

In addition to Eq. (5), the collapsing force can be also written in the 'potential' form in 'momentum space' with the Higgs-type 'collapsing potential': $F_c = -\partial V_c/\partial p, \quad V_c = (g/2)(q^2 - p^2)^2$ (Fig.1). This illustrates the remarkable link between our model and the phenomena of symmetry breaking, phase transitions, and synergetics. The collapsing potential can be compared to the 'mean field' and the measured quantity—to the 'order parameter'.

Solving Eq. (10) with the initial condition (3), yields

$$p(x,t) = \frac{qp_0(x)B(t)}{\sqrt{p_0^2(x)[B^2(t) - 1] + q^2}}, \tag{11}$$

$$B(t) = \exp{(gq^2 t)}, \quad p_0(x) = iq \tan kx. \tag{12}$$

Finally, using the definition (2) we can determine the *continuous* behavior of the wave function *all the way through the measuring process*—in other words, *the process of the wave function collapse*:

$$\psi(x,t) = \frac{B(t)\cos kx + \sqrt{1 - B^2(t)\sin^2 kx}}{B(t) + 1} N(t), \tag{13}$$

where $N(t)$ is the normalization factor. In particular, we obtain the following formula for the characteristic timescale τ_c of the collapse: $\tau_c \sim 1/(gq^2)$.

Asymptotically, when $B(t) \gg 1$, the wave function (13) can take one of the two forms (in each domain of length $\pi/2k$):

$$\psi_+ = e^{ikx} \quad \text{or} \quad \psi_- = e^{-ikx}. \tag{14}$$

This two-valuedness arises due to the two-valuedness of the square roots in Eqs. (11,13). An interesting question now is: How does the wave function 'know' which way to collapse: ψ_+ or ψ_-? The answer is that if the initial wave function has *the exact* form (3), then *it does not know,* i.e., the two-valued ambiguity in (14) *cannot be resolved.*

This is because the initial p_0 is *purely imaginary.* If, on the other hand, p_0 has a non-zero real part $Re\ p_0$, then the definite outcome arises: in case of positive real part, $Re\ p_0 > 0$, the asymptotic wave function is $\psi_+ = e^{ikx}$; if the real part is negative, $Re\ p_0 < 0$, the asymptotic wave function is $\psi_- = e^{-ikx}$. Thus the measurement outcome is completely determined by the sign of the real part of p_0. It is natural to presume that $Re\ p_0$ can take positive and negative values with equal probability as required by the Born rule.

We emphasize that only *infinitesimal* non-zero values of $Re\ p_0$ are required for the emergence of a definite outcome in the scheme based on the simple equation (10).

A question may be raised whether $Re\ p_0$ can be considered a hidden variable, the dependence on which yields a step function.

Using a step function to represent the outcome dependence on $Re\ p_0$ leads to the following problems: If the dependence of the outcome on $Re\ p_0$ is represented by a step function, the domain of this function should be infinitesimally small. If finite values of $Re\ p_0$ are allowed to be included in that domain, then these values would be incompatible with the state described by the wave function (3). But if we consider other states compatible with finite values of $Re\ p_0$, then such a step function would not be a correct predictor of the outcome. Therefore, the outcome dependence on $Re\ p_0$ has a more singular character than the step function. By the same argument, the outcome dependence on $Re\ p_0$ cannot be represented by *any* function with finite domain. Also, it cannot be represented by a discrete binary variable λ taking on values ± 1 because in that case the dependence of this 'stand alone' variable on $Re\ p_0$ would be lost.

For these reasons we prefer not to call $Re\ p_0$ a hidden variable.

The real issue, though, is not so much in the name but in a possible contradiction with various no-go theorems which have been formulated in relation to theories with hidden variables of various sorts. These issues are discussed in Sec. V.F where it is shown that no contradictions arise.

3. Interpretation

A few remarks are in order concerning the interpretation of our 'collapsible equation', Eq. (10).

(a) It is not supposed to be a fundamental equation describing the collapse process exactly in every detail. Rather, it is a phenomenological, effective approximation which is meant to capture the main features of the collapse: instability, irreversibility, and symmetry breaking, which could lead to observable consequences such as the finite time-scale of the collapse process.

In any case, the construction of more elaborate models of the wave function collapse seems to require more experimental data than is available at the moment.

(b) In contrast with the Bohm-Bub models of explicit collapse [20], Eq. (10) does *not* contain hidden variables.

A large number of works *replace* the Schrödinger equation with a non-linear one (see, e.g., [22, 23]). In contrast, our collapsible equations *complement* the Schrödinger equation. They have a different mathematical structure and distinct observational consequences.

(c) The "superluminality" argument which is often invoked to limit the use of non-linear equations in quantum mechanics does *not* apply in our case for the reasons discussed in Sec. V.

4. Experimental Consequences

As with any phenomenological equation, care is required in comparing its predictions with experiment. Although our derivation allows one to establish some necessary conditions of its validity, these conditions may not be sufficient, and only experiment can decide on this issue. First of all, a certain balance between the characteristic times involved should be observed. Ideally, one would aim at the interaction-switching time much less than the collapse time, and the collapse time much less than the 'energy' time-scale $\hbar/\tau_E$ where E is the characteristic energy scale involved in a particular experiment.

One scheme of an experiment could be as follows. The initial state is created, then the measurement interaction is switched on at the instant $t = 0$, then the measurement result is read off. In the conventional picture, the interval between the second and third events can be as small as possible (it is bounded from below by the resolution time of the detector). However, in our picture, the collapse time cannot be less than τ_c. Therefore, if the time-scale τ_c exceeds the resolution time-scale, the theory can be tested.

To compare the sensitivities of different kinds of experiments, it is convenient to introduce the dimensionless collapse time $\kappa = \tau_c/\tau_E$. Except for the pioneering work of Papaliolios [25] back in 1967, no dedicated searches for the finite time of the collapse have been performed (to my knowledge). So we attempt to obtain rough estimates on the allowed values of τ_c and κ from various other experimental results, assuming that null effects were observed in them on a time scale of τ_m, which is the shortest time scale probed in a specific experiment (Table I).

Table 1. Experimental constraints on the collapse time τ_c.

Experiment	Ref.	τ_m (s)	τ_E (s)	r
Photon polarization	[25]	7.5×10^{-14}	$\sim 10^{-15}$	$\sim 10^{2}$
Neutron interferometry	[26]	2.7×10^{-2}	2.3×10^{-12}	10^{10}
Quantum jumps	[27]	1	$\sim 10^{-15}$	$\sim 10^{15}$
Nonlinearity test	[28]	1	$\sim 10^{-9}$	$\sim 10^{9}$
Femtosecond optics	[29]	$\sim 10^{-13}$	$\sim 7 \times 10^{-17}$	$\sim 10^{3}$
Bose-Einstein condensate	[30]	$\sim 10^{-4}$	1.8×10^{-3}	$\gtrsim 5 \times 10^{-2}$
EPR correlations	[31]	5×10^{-12}	$\sim 10^{-15}$	5×10^{3}

Absolute upper bound: $\tau_c \lesssim \tau_m$. Relative upper bound: $\tau_c/\tau_E \lesssim r$. All figures are to be taken as order-of-magnitude estimates rather than rigorous limits.

It is natural to start with the experiments [26, 28] that tested specific non-linear versions of the Schrödinger equation [22, 23, 24]. However, they turn out *not* to be the most sensitive for our purposes. The well-known observation of 'quantum jumps' [27] is another natural place to look for constraints, but again they are found to be rather weak. Next, we turn to the ultrafast measurements using femtosecond pulses of laser light [29].

Here, the limits begin to improve. Finally, an excellent source of constraints appears to be experiments with Bose-Einstein condensates. The observed dynamics turned out surprising in many respects, and it would be interesting to see if the idea of a finite-time collapse could be helpful there.

This area is virtually unexplored, and not one, but many promising directions of attack exist. Consequently, without attempting to be exhaustive, we restrict ourselves to outlining just two general strategies.

One approach is to look for the longest characteristic times τ_E, comparable to or better than those achieved in the BEC experiments. For example, the distances between the energy levels of ultra-cold neutrons bouncing from a horizontal mirror in the earth gravity field are of the order of 10^{-12} eV (or peV) [33] which corresponds to $\tau_E \sim 10^{-3}$ s. Because times, such as the time of flight, can be measured with much better accuracy, this type of experiment (suggested for other purposes) could be a useful starting point for a test proposal.

An alternative strategy would be to start from the experiments with the shortest possible times τ_m, such as those involving ultrashort pulses. This is a rapidly evolving area progressing from the femtosecond scale to atto-, zepto-, and even yoctoseconds (10^{-24} s).

We emphasize that if the theory is confirmed, that would not at all mean that the conventional quantum mechanics is refuted. It would only mean that the approximation of the instantaneous collapse is too crude and should be replaced by a more realistic approximation of a finite time-collapse. As was mentioned in the introduction, the assumption of instantaneous collapse is just an approximation rather than a fundamental principle of quantum theory, and the validity of this approximation does not affect the validity of quantum theory as a whole.

Similarly, if our specific Eq. (10) is experimentally excluded, that would not mean that the whole approach is invalidated, because a variety of different equations could be built on the same conceptual basis, which could lead to different experimental predictions.

5. Discussion

In constructing the collapse-enabled Schrödinger equations, our method is based on three well-established principles:

1. *Symmetry breaking.* The binary symmetric measurements on one particle in the absence of decoherence are the fundamental prototype measurements that should be studied first, both theoretically and experimentally.

Furthermore, the collapsible equations for such measurements should be constructed phenomenologically using the minimal set of most general assumptions. Specifically, we single out the following two conditions that are widely accepted as reliable, non-controversial characteristics of the collapse process:

2. *Irreversibility.* This is generally acknowledged as the most important, essential feature of the collapse.

3. *Instability* (or amplification). The key role of these principles in the theory of measurement was succinctly formulated by Bohr as follows:"...every atomic phenomenon is closed in the sense that its observation is based on registrations obtained by means of suitable *amplification* devices with *irreversible* functioning such as, for example, permanent marks on the photographic plate caused by the penetration of electrons into the emulsion." (italics added—A.I.). In the earlier literature, the same principles were often described in the language of "uncontrollable perturbations" arising when the quantum system interacts with the measuring device.

We note that the requirement of finite time is *not* among our basic principles. Rather, it comes out as a natural consequence of them.

Our method reflects the fact that decades of concerted efforts toward clarifying the fundamental mechanisms of collapse and the measurement process have met with limited success. This means that the collapse problem is an extremely challenging as well as a highly interdisciplinary one. It involves several distinct areas of theoretical physics and modern mathematics. As a consequence, there is little hope of progress unless one uses the phenomenological approach first and asks questions about first principles later.

Wave function collapse can be regarded as a 'macroscopic quantum phenomenon'. Therefore we can turn for inspiration to the successful theories of other macroscopic quantum phenomena, such as superfluidity and superconductivity.

In particular, it would be interesting here to draw a parallel with the puzzle of superconductivity, which took almost 50 years to unravel and is still at the forefront of physics research. This ultimate success was based on the consistent use of the phenomenological, effective-theory type of approaches before embarking on the final quest for a microscopic theory. The same strategy could be worth trying in the collapse problem.

The analogy can be continued in another respect. The phenomenological theory of superconductivity of Ginzburg and Landau uses a non-linear equation for the wave function. This does not mean, however, that the superposition principle is violated. Likewise, our use of a non-linear collapsing equation does not necessarily mean the violation of that principle because it is meant to be understood as a description at the phenomenological level, which is not necessarily fundamental.

Many alternative models can be constructed by choosing different representations, forms of the collapsing force and/or adding hidden variables to the fundamental

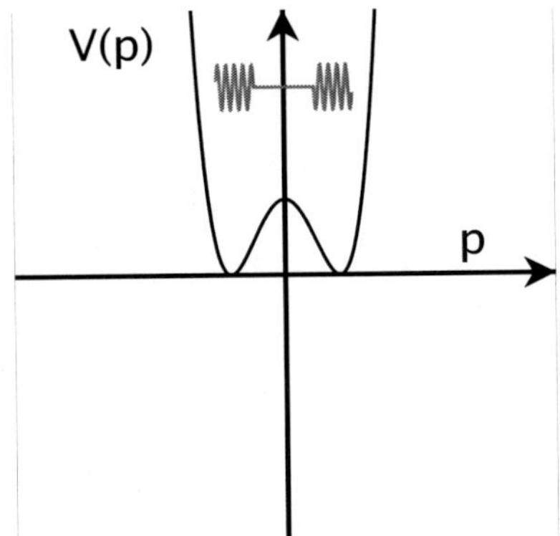

Figure 2. The initial wave function is a symmetric superposition of opposite momentum values $p = +q$ and $p = -q$.

general equations belonging to the family exemplified by Eqs. (4) and (10). However, given the probable complexity of the collapse phenomenon, it would be desirable to focus first (both theoretically and experimentally) on the most fundamental, the "purest" form of collapse—the binary symmetric measurement.

With suitable modifications, our methods could be applied in other subfields of quantum theory where a non-potential generalization of the Schrödinger equation is required by the conditions of the problem. Examples include the theory of dissipative systems, motion under the forces of radiative friction, etc. More generally, one can expect that interesting connections with nanoscience, quantum gravity and quantum cosmology could also be possible.

5.1. Schematic of Main Finding

Quantum state collapse is represented as a kinetic process of irreversible symmetry breaking driven by the instability of the initial state (Fig. 1-3).

5.2. Physical Ideas Underlying Our Framework

Our approach and results may look deceptively simple, and questions could be raised as to what makes this possible.

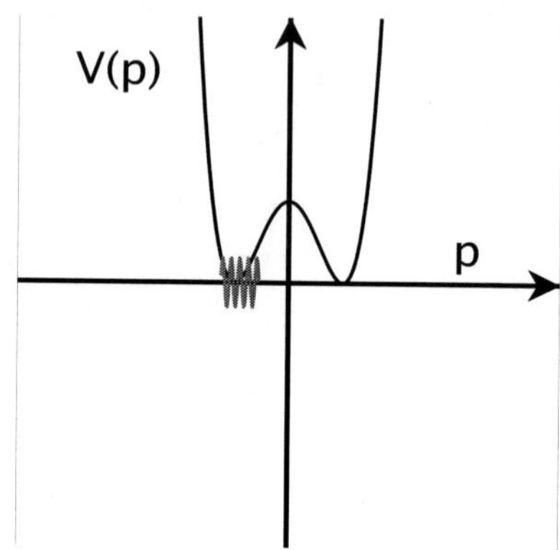

Figure 3. After the measurement interaction gave the result $p = -q$, the wave function collapses to $\psi_- = exp(-iqx)$.

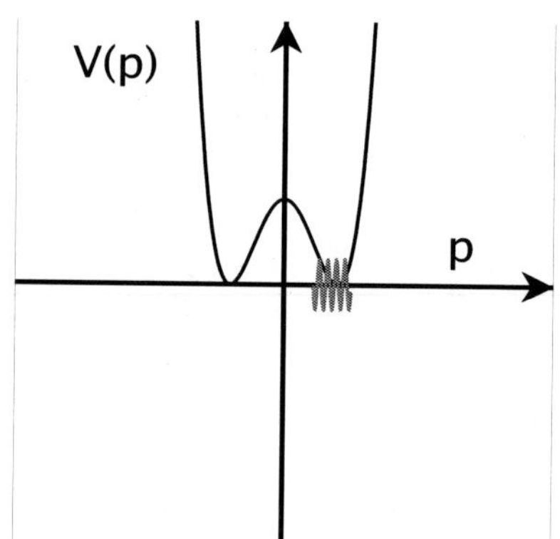

Figure 4. After the measurement interaction gave the result $p = +q$, the wave function collapses to $\psi_+ = exp(+iqx)$.

The explanation is that all the hard work is done by combining six powerful insights from several fronts

of theoretical physics, which have not been put together before:

-complex quantum Hamilton-Jacobi formulation of quantum mechanics

-principle of symmetry breaking

-the effective theory approach

-the idea of irreversibility

-the measured quantity as the order parameter

-Bohr's insight that quantum phenomena occur due to irreversible amplifying interaction of classical and quantum systems.

5.3. Where Do We Stand in Terms of Various Interpretations?

Although our formulation is designed to be interpretation-neutral, it can also be seen as the logical development of the conventional Copenhagen interpretation. As discussed in the main text, those who favour the statistical ensemble interpretation of quantum mechanics would probably be skeptical about our approach. In the spirit of "experimental metaphysics", we believe that experimental tests may help to clarify or update one's philosophical position.

The same can be said about the Many-Worlds interpretation. In fact, experimental searches for collapse dynamics are needed from the standpoint of *any* interpretation because the results of such experiments will help to decide which one is closer to Nature.

If these searches fail to produce evidence for collapse, this would be a serious argument in favour of the collapse-denying interpretations, including Many-Worlds and statistical. Such arguments are hard to achieve from just theoretical reasoning.

Also, it is well-known that in some situations the wave function is 'more real' than in others (superconductivity, Bose-Einstein condensation, etc.). Consequently, collapse may be a more real phenomenon in those situations. In any case, the wave function satisfying our collapsible equation should be interpreted as an 'effective' or 'interpolating' wave function rather than a 'genuine' wave function which ceases to exist once the system starts to interact with the measuring device, according to the standard (von Neumann's) model of the measurement process.

5.4. Comparison with the Approaches of Bohr and von Neumann

In any treatment of the measurement problem the question arises: is the measuring apparatus to be treated as a quantum system or not? On this crucial point, the opinions of quantum theorists appear to be split. One school of thought (Bohr, Heisenberg, Dirac, and Born) accepts Bohr's idea that the measuring device should be described approximately as classical and not quantum-mechanical. Consequently, the wave function of the apparatus is not introduced and the Schrödinger equation for the combined set "system plus apparatus" is not discussed or solved. The alternative school (von Neumann, Landau and Peierls, Pauli, and most modern textbooks) assumes that the measuring device should be treated on the same footing as any other quantum system. As a consequence, most modern treatments introduce the wave function of the apparatus as a matter of fact that does not require further justification.

At the moment, one of the fruitful if challenging issues in quantum mechanics is the question whether the superposition principle is universally valid or not. Obviously, this is just another way to ask the same question. Again, this problem is attacked not by purely philosophical discussion, but in the spirit of 'experimental metaphysics'. However, despite impressive progress in the area, it would be fair to say that the final solution is not yet on the horizon.

We do not explicitly introduce the wave function of the measuring device. This is in accord with Bohr's ideas that the the measuring device is best described approximately and classically rather than quantum-mechanically. In fact, Bohr appears to have reservations regarding the crucial step of "quantizing the apparatus" (first made by von Neumann) as he did not discuss it in his published works. From the modern perspective, this may be viewed as a flaw or even a serious defect of such an interpretation. On the other hand, Bohr's reservation might probably be justified because to date, efforts spent on quantizing the apparatus have not been completely successful. Moreover, this problem is considered fundamental but intractable in the forseeable future. For example, it was pointed out [5] that no observer can write down his own wave function just because human brain capacity is not large enough for that. A similar type of argument applies for an inanimate device as well.

We, of course, do not pretend to solve this part of the problem. Instead, we go around it and concentrate on the behaviour of the quantum system rather than the measuring device. In this way, we act in the spirit of 'effective' theory, which is well-known and extremely fruitful in the area of quantum fields (see, e.g., [45]), but is less appreciated in the context of quantum mechanics. The essence of the effective theory approach is to integrate out all unknown degrees of freedom and focus on what is essential and observable.

5.5. Relation to Decoherence

Although the theory of decoherence made an important step towards solving the collapse problem, it has not solved it.

Our paper does not include decoherence explicitly, but this is not because it is irrelevant. The real reason is that the collapse puzzle is a difficult and unsolved task. The common approach of a theoretical physicist in such cases is to first tackle an idealized setting where decoherence is ignored.

Moreover, this simple setting could be much more realistic than it seems at first sight. Indeed, we could imagine physical conditions under which the effects of decoherence are greatly reduced, such as very low noise, low temperature and similar types of environment. In this 'very clean' environment we expect that decoherence will be greatly affected, but the collapse will not. Consequently, analysis of collapse in the absence of decoherence is not merely an academic exercise, but a viable physical model for specific environmental conditions.

5.6. The Einstein-Podolsky-Rosen Set-up and No-go Theorems

In this paper, we do not treat any EPR-related issues leaving them for further work. We feel that the phenomenon of collapse, if real, could be complicated and not necessarily universal for all types of systems and experiments. One may even wonder if the quantum collapse could be as complex as its gravitational cousin. In any case, it seems prudent to approach this difficult problem in a step-by-step manner starting with the simplest possible set-up: a measurement made on one non-relativistic particle.

As Braginsky and Khalili note in Ref. [5] (p. 28): "The presence of two or more degrees of freedom can change the character of the measurement substantially".

Perhaps, to emphasise the differences, it could be helpful to introduce special terms like, for example, "contact" and "non-contact" collapse, the latter describing the EPR-type situations.

We also note that under a different set of assumptions, an EPR-type experimental search for a finite-time collapse was recently proposed in Ref. [35].

Because of the fact that our simple model (10) treats only the case of one particle and two measurement outcomes, it avoids tensions with Bell's inequalities [36] as well as various other no-go theorems due to Gleason [37]; Kochen and Specker [38]; Conway and Kochen (or "the Free Will theorem") [39]; Colbeck and Renner [40]; Pusey, Barrett and Rudolph [3].

More specifically, both Gleason's and Kochen-Specker's theorems require that the dimensionality of Hilbert space be greater than two [37, 38]—which does not apply in our case (10).

The Free Will theorem, in both original and strengthened formulations, requires a two-particle state due to its TWIN axiom [39]. Therefore, it does not clash with our model (10) either.

Colbeck and Renner [39] extended Bell's prohibition of local hidden variables by considering hidden variables that have both local and global components and ruling them out as well. Again, our model (10) is immune from this exclusion theorem because it (the theorem) requires at least a two-particle system, among other things [40].

Pusey, Barrett and Rudolph [3] have recently proved another no-go theorem excluding a certain class of theories with hidden parameters λ. The following quote from Ref. [3] summarizes their work nicely: "Our main result is that for distinct quantum states $|\psi_0\rangle$ and $|\psi_1\rangle$, if the distributions $\mu_0(\lambda)$ and $\mu_1(\lambda)$ overlap (more precisely: if Δ, the intersection of their supports, has non-zero measure) then there is a contradiction with the predictions of quantum theory."

It follows immediately that this result does not apply to our model (10) because for distinct quantum states the distributions of our infinitesimal parameters do not overlap.

5.7. Non-linear Quantum Mechanics and Superluminality

It has been proposed by Gisin [41], among others, that non-linear modifications of quantum mechanics could lead to faster than light communication. However, several points have to be taken into account here:

1) Our collapsible Schrödinger equations are *not* of the type considered in Ref. [41]. Our non-linearity arises *only* at the measurement stage, while equations studied in Ref. [41] are non-linear *always*. For this reason, superluminal communication is not made possible by our equations and the objection does not apply.

2) For completeness, we also mention that in subsequent papers [42, 43, 44] it was pointed out that the argument of Ref.[41] is not free from possible loopholes. However, due to the above point, we do not need to appeal to these loopholes in order to demonstrate that our equations are free from the superluminality objection.

5.8. Is Consciousness Needed for Collapse?

In our approach, an observer or the consciousness is not required for collapse to occur (cf. [46]). Moreover, our experimental upper limits on the collapse time (Table 1) show that this process is too fast for human consciousness which typically operates on much longer time-scales.

5.9. Collapse and Relativity

Relativistic generalisations of the collapse concept are notoriously difficult and are not discussed here. We would like to stress, however, that at the moment the proposed collapse mechanism does not appear to be incompatible with relativity. On the contrary, a relativistic generalisation of our collapsible equations seems a worthwhile direction to pursue.

5.10. de Broglie-Bohm Formulation

Some readers may get an impression that we are using or advocating some version of the de Broglie-Bohm (dBB) interpretation of quantum mechanics [47]. We emphasise that this is *not* the case[2]. What we do is just borrow a useful form of rewriting the Schrödinger equation. This can be considered as a purely mathematical transformation that is completely neutral with regard to interpretational issues. The dBB interpretation involves much more than just a rewriting, e.g., particle trajectories. These extra features are not necessary in our approach.

5.11. Other Types of Measurements

Today we know many more types of measurements than in 1930s. Examples include weak, indirect, non-demolition, continuous, interrupted and other varieties

---

[2]This phrase should not be construed as an implicit criticism of such interpretations.

of measurements. We emphasise that all these are beyond the scope of the present paper.

Our focus here is only on the simplest, "canonical" kind of measurement, which was also historically one of the first examples where the measurement problem was born. Without a thorough understanding of this example, there is little chance of progress in the more intricate cases.

Moreover, it would be perhaps over-optimistic to expect that one and the same theory can describe *all* kinds of measurements. (It would be similar to expecting, for instance, that one and the same theory can describe both strong and gravitational interactions.)

5.12. Theories of Dynamical Collapse

On a first view, our framework can be seen as a competitor to the well-known theories that represent collapse as a continuous dynamical process [48]. In fact, it is not. The main reason is that the two approaches set different goals from the start even if the term "collapse" happens to figure prominently in both of them.

Briefly, this term is used in two rather different senses so one wonders if two different terms would be more appropriate here. Our collapse has been induced by the measurement ("M-collapse"). In the theories of dynamical collapse, it occurs spontaneously, during free evolution ("S-collapse"). This is a considerable difference with far-reaching consequences.

The theories of dynamical collapse do not address the mechanism of M-collapse. Likewise, we do not discuss the S-collapse. We ask the question: what happens to the wave function during the measurement? The theories of dynamical collapse ask: Why do the wave functions not spread out so much that the classical picture becomes blurred (problem of macro-objectification)?

It might be possible that in the end, the answers to both questions could come from one and the same ultimate source. However, for all similarities between the two problems there seem to be as many differences, so it would be reasonable to treat them as separate issues for the time being.

Our model and the dynamical collapse theories are made for two different purposes; they have different mathematical structure and different observational consequences.

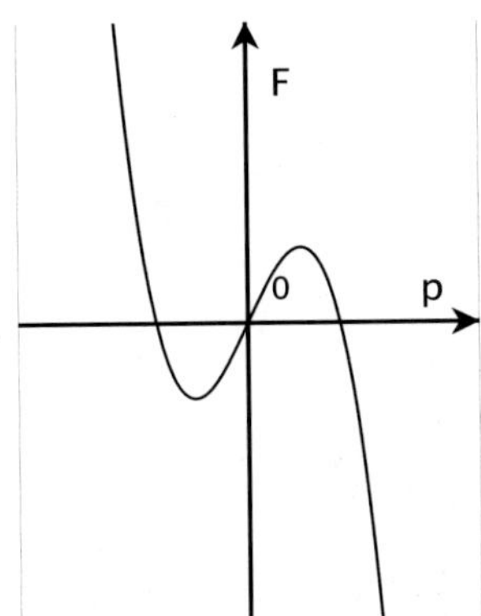

Figure 5. The schematic form of the 'collapsing force' F(p). This 2-dimensional plot assumes that p and therefore F are real numbers. (In general, both are complex so a 4-dimensional plot would be required.)

5.13. Meaning of collapsing force

Introduction of the collapsing force (Fig. 4) is the main phenomenological element of our program. As usual in such cases, we do not attempt to derive it from first principles, but rather postulate it on the combined strength of a series of independent arguments:

—The irreversible collapse during a measurement can be viewed as sort of relaxation process. Then the appearance of a collapsing force is naturally understood as the emergence of a relaxation-driving dissipative force without which relaxation would be impossible.

—In the same vein, the kinetic mechanism of collapse invites a parallel with classical kinetic theory. The collapsing force, then, plays a similar role to the collision term in the Boltzmann equation.

—The task of constructing the collapsible Schrödinger equation can be formulated as the problem of finding a *smooth interpolation* between the uncollapsed and collapsed wave functions. A new term that is needed for such an interpolation would be equivalent to a collapsing force.

—What arguments force us to assume that the collapse occurs instantaneously, given that we are not aware of any other instantaneous processes in Nature?

—If, on the other hand, the collapse is not instantaneous but takes a finite time, then it seems plausible for purely mathematical reasons that one or another kind of extra terms ('generalised collapsing forces') should be introduced into the CQHJ-Schrödinger equation to produce this finite time.

—Within the CQHJ approach, the Schrödinger equation acquires a form so similar to Newton's second law that it almost 'begs for' introducing such a new force and exploring its consequences.

5.14. Is Quantum Mechanics to Be Modified?

Do our equations require a modification of standard quantum mechanics? At first sight, the answer to this would be "Yes", but we believe that such a conclusion would be premature and unwarranted.

As was mentioned in the main text, our goal was not to modify quantum mechanics, but to obtain effective equations for the collapse process. The fact that our equations are non-linear, and the Schrödinger equation is linear is not enough, in our view, to claim that these equations violate quantum mechanics.

The reason is that the Schrödinger equation is exact, and our equation is approximate. But it is well-known that the exact and linear equation (Schrödinger) can lead to approximate non-linear ones when suitable approximations are made (equations of Hartree-Fock, Gross-Pitaevsky, etc.).

Indeed, as emphasized by Fock [49], in quantum theory the distinction between "fundamental" and "effective" concepts becomes a matter of convention because approximations play the principal, not a secondary role. For example, from the perspective of quantum field theory, the Schrödinger equation itself becomes an approximate "effective" equation, rather than a "fundamental" law. Finally, from a purely experimental perspective, the 'fundamental or effective' question is hardly the most pressing one.

6. Conclusions and Outlook

To sum up, a new class of quantum-mechanical non-linear equations for the wave function collapse has been proposed that combines conceptual simplicity and predictive power. Also, we have reported a set of constraints and outlined future opportunities for sensitive experimental tests.

It is exciting to realize that only two or three orders of magnitude separate us from deeper, dedicated probes of the wave function collapse. The results will help to clarify both foundational and practical issues in quantum mechanics, such as the status of the Many-Worlds interpretation and the ultimate limits on the speed of quantum computation and information processing.

Acknowledgments

The author thanks M. Albrow, R. Dörner, L. Ignatieva, V. A. Kuzmin, J.-P. Magnot, M. E. Shaposhnikov, and F. Thaheld for taking time for reading, valuable comments and discussions.

Appendix

Complex Quantum Hamilton-Jacobi Equation

To make the article self-contained, we provide below a derivation of the complex quantum Hamilton-Jacobi equation.

Let us start with the ordinary Schrödinger equation for a particle of mass m in the potential $V(\mathbf{r})$:

$$i\hbar\psi_t = -\frac{\hbar^2}{2m}\Delta\psi + V\psi, \tag{15}$$

and introduce a new variable $\mathbf{p}(\mathbf{r},t)$:

$$\mathbf{p} = \frac{\hbar}{i}\frac{\nabla\psi}{\psi}. \tag{16}$$

Our goal is to obtain a closed dynamical equation for $\mathbf{p}$. Differentiating Eq. (16) with respect to time, we find:

$$\mathbf{p}_t = \frac{\hbar}{i}\frac{\nabla\psi_t}{\psi} - \mathbf{p}\frac{\psi_t}{\psi}. \tag{17}$$

It is convenient to start ψ-elimination from the second term of Eq. (17). For this purpose, we first rewrite $\Delta\psi$ as

$$\Delta\psi = \nabla\left(\frac{i}{\hbar}\mathbf{p}\psi\right) = \frac{i}{\hbar}\left(\psi\nabla\cdot\mathbf{p} + \mathbf{p}\nabla\psi\right). \tag{18}$$

Inserting this into the Schrödinger equation (15) and multiplying both sides by $i\mathbf{p}/(\hbar\psi)$, we obtain:

$$-\mathbf{p}\frac{\psi_t}{\psi} = \frac{1}{2m}\mathbf{p}\nabla\cdot\mathbf{p} + \frac{i}{2m\hbar}\mathbf{p}p^2 + \frac{i}{\hbar}\mathbf{p}V. \tag{19}$$

To eliminate ψ from the first term of Eq. (17), we substitute the right-hand side of Eq. (18) into the Schrödinger equation (15) instead of $\Delta\psi$, then take the gradient of both parts and divide both of them by ψ, which yields:

$$-i\hbar\frac{\nabla\psi_t}{\psi} = \frac{i\hbar}{2m}\nabla(\nabla\cdot\mathbf{p}) - \frac{1}{2m}\mathbf{p}\nabla\cdot\mathbf{p} - \frac{1}{2m}\nabla p^2 - \frac{i}{2\hbar m}p^2\mathbf{p} - \nabla V - \frac{i}{\hbar}\mathbf{p}V. \tag{20}$$

We now add this formula, part-by-part, to Eq. (19) and then use Eq. (17) to finally obtain:

$$\mathbf{p}_t = -\nabla V - \frac{1}{2m}\nabla p^2 + \frac{i\hbar}{2m}\nabla(\nabla\cdot\mathbf{p}). \tag{21}$$

So the complete elimination of the wave function is indeed possible, and Eq. (21) reproduces the dynamical equation that plays the role of the Schrödinger equation in the CQHJ approach. A slight rearrangement gives it an elegant shape:

$$\mathbf{p}_t = -\nabla H, \tag{22}$$

$$\hat{\mathbf{p}} = -i\hbar\nabla, \quad H = V + \frac{p^2}{2m} + \frac{\hat{\mathbf{p}}\mathbf{p}}{2m}. \tag{23}$$

References

[1] Leggett, A. J. The quantum measurement problem. *Science* **307**, 871-872 (2005)

[2] von Neumann, J. Mathematische Grundlagen der Quantenmechanik. *(Julius Springer-Verlag, Berlin, 1932)* [English transl.: *Princeton University Press, Princeton, N.J., 1955*].

[3] Pusey, M. F., Barrett, J. & Rudolph, T. On the reality of the quantum state. *Nature Physics* **8**, 476-479 (2012).

[4] Nielsen, M. A. & Chuang, I.L. Quantum computation and quantum information. *(Cambridge University Press, Cambridge, 2010).*

[5] Braginsky, V. B. & Khalili, F. Ya. Quantum Measurement. *(Cambridge University Press, Cambridge, 1992).*

[6] Adler, S. Why decoherence has not solved the measurement problem. *Stud. Hist. & Phil. Sci.* **34**, 135-142 (2003).

[7] Schlosshauer, M., Kofler, J. and Zeilinger, A. A Snapshot of Foundational Attitudes Toward Quantum Mechanics. arXiv:1301.1069 [quant-ph].

[8] Ghirardi, G.C., Rimini, A., & Weber, T. Unified dynamics for microscopic and macroscopic systems. *Phys. Rev. D* **34**, 470-491 (1986).

[9] Pearle, P. Combining stochastic dynamical state-vector reduction with spontaneous localization. *Phys. Rev. A* **39**, 2277-2289 (1989).

[10] Weinberg, S. Collapse of the state vector. *Phys. Rev. A* **85**, 062116 (2012).

[11] Penrose, R. On gravity's role in quantum state reduction. *Gen. Rel. Grav.* **28**, 581-609 (1996).

[12] Diósi, L. The frictional Schrödinger-Newton equation in models of wave function collapse. *J. Phys.: Conf. Ser.* **67**, 012024 (2007).

[13] Leacock, R. A. & Padgett, M. J. Hamilton-Jacobi theory and the quantum action variable. *Phys. Rev. Lett.* **50**, 3-6 (1983).

[14] John, M. V. Modified de Broglie-Bohm approach to quantum mechanics. *Found. Phys. Lett.* **15**, 329-343 (2002).

[15] Yang, C.-D. Quantum dynamics of hydrogen atom in complex space. *Ann. Phys.* **319**, 399-443 (2005).

[16] Goldfarb, Y., Degani, I. & Tannor, D. J. Bohmian mechanics with complex action: A new trajectory-based formulation of quantum mechanics. *J. Chem. Phys.* **125**, 231103 (2006).

[17] Poirier, B. Flux continuity and probability conservation in complexified Bohmian mechanics. *Phys. Rev. A* **77**, 022114 (2008).

[18] Chou, C. C., Sanz, Á. S., Miret-Artés, S. & Wyatt, R. E. Hydrodynamic view of wave-packet interference: quantum caves. *Phys. Rev. Lett.* **102**, 250401 (2009).

[19] Ignatiev, A. Yu. How fast is the wave function collapse? arXiv:1204.3373 [quant-ph]; *J. Phys.: Conf. Ser.* **410**, 012153 (2013).

[20] Bohm, D. & Bub, J. A proposed solution of the measurement problem in quantum mechanics by a hidden variable theory. *Rev. Mod. Phys.* **38**, 453-459 (1966).

[21] Pearle, P. M. Reduction of the state vector by a nonlinear Schrödinger equation. *Phys. Rev. D* **13**, 857-868 (1976).

[22] Bialynicki-Birula, I. & J. Mycielski, Nonlinear wave mechanics. *Ann. Phys.* **100**, 62-93 (1976).

[23] Weinberg, S. Precision tests of quantum mechanics. *Phys. Rev. Lett.* **62**, 485-488 (1989).

[24] Mizrahi, S., Dodonov, V.V. and Otero, D. Nonlinear Schrödinger-Liouville equation with antihermitian terms. *Phys. Scr.* **57**, 24 (1998).

[25] Papaliolios, C. Experimental test of a hidden-variable quantum theory. *Phys. Rev. Lett.* **18**, 622-625 (1967).

[26] Gähler, R., Klein, A. G. & Zeilinger, A. Neutron optical tests of nonlinear wave mechanics. *Phys. Rev. A* **23**, 1611-1617 (1981).

[27] Nagourney, W., Sandberg, J. & Dehmelt, H. Shelved optical electron amplifier: observation of quantum jumps. *Phys. Rev. Lett.* **56**, 2797-2788 (1986).

[28] Bollinger, J. J. *et al.* Test of the linearity of quantum mechanics by rf spectroscopy of the ^{9}Be$^+$ ground state. *Phys. Rev. Lett.* **63**, 1031-1034 (1989).

[29] Bauer, M. *et al.* Direct observation of surface chemistry using ultrafast soft-X-ray pulses. *Phys. Rev. Lett.* **87**, 025501 (2001).

[30] Donley, E. A. *et al.* Dynamics of collapsing and exploding Bose-Einstein condensates. *Nature* **412**, 295-299 (2001).

[31] Zbinden, H. Brendel, J., Gisin, N. & Tittel, W. Experimental test of non-local quantum correlation in relativistic congurations. *Phys. Rev. A* **63**, 022111 (2001).

[32] Jenke, T. *et al.* Realization of a gravity-resonance-spectroscopy technique. *Nature Physics* **7**, 468-472 (2011).

[33] Durstberger-Rennhofer, K., Jenke, T. & Abele, H. Probing the neutrons electric neutrality with Ramsey spectroscopy of gravitational quantum states of ultracold neutrons. *Phys. Rev. D* **84**, 036004 (2011).

[34] Plonka-Spehr, C. *et al.* An optical device for ultra-cold neutrons—Investigation of systematic effects and applications. *Nucl. Instrum. & Methods A* **618**, 239-247 (2010).

[35] Moreno, M. G. M. and Parisio, F. Investigation of the collapse of quantum states using entangled photons. *Phys. Rev. A* **88**, 012118 (2013).

[36] Bell, J.S. On the Einstein-Podolsky-Rosen paradox. *Physics* **1**, 195 (1964); Brunner, N., Cavalcanti, D., Pironio, S., Scarani, V. and Wehner, S. Bell nonlocality. arXiv: 1303.2849 [quant-ph].

[37] Gleason, A. M. Measures on the closed subspaces of a Hilbert space. *J. Math. Mech.* **6**, 885 (1957).

[38] Kochen, S. and Specker, E. J. The Problem of Hidden Variables in Quantum Mechanics. *J. Math. Mech.* **17**, 59 (1967).

[39] Conway J. and Kochen, S. The Free Will Theorem. *Found. Phys.* **36**, 1441 (2006); arXiv: 0807.3286 [quant-ph].

[40] Colbeck, R. and Renner, R. Hidden Variable Models for Quantum Theory Cannot Have Any Local Part. *Phys. Rev. Lett.* **101**, 050403 (2008).

[41] Gisin, N. Stochastic quantum dynamics and relativity. *Helv. Phys. Acta* **62**, 363-371 (1989).

[42] Ferrero, M., Salgado, D. & Sánchez-Gómez, J.L. Nonlinear quantum evolution does not imply supraluminal communication. *Phys. Rev. A* **70**, 014101 (2004).

[43] Kent, A. Nonlinearity without superluminality. *Phys. Rev. A* **72**, 012108 (2005).

[44] Brody, D.C., Gustavsson, A.C.T. & Hughston, L.P. Nonlinearity and constrained quantum motion. *J. Phys. A: Math. Theor.* **43**, 082003 (2010).

[45] Weinberg, S. The Quantum Theory of Fields, v. 1. *(Cambridge University Press, Cambridge, 1995).*

[46] Thaheld, F. H. A modified approach to the measurement problem: Objective reduction in the retinal molecule *prior* to conformational change. *BioSystems* **92**, 114 (2008).

[47] Holland, P.R. The Quantum Theory of Motion. *(Cambridge University Press, Cambridge, 1993).*

[48] Bassi, A., Lochan, K., Satin, S., Singh, T.P., and Ulbricht, H. Models of wave-function collapse, underlying theories, and experimental tests. *Rev. Mod. Phys.* **85**, 471 (2013).

[49] Fock, V. A. The principal role of approximate methods in theoretical physics. *Usp. Fiz. Nauk* **16**, 1070 (1936).

Chapter 4

The Minkowski World

B. P. Kosyakov *
Russian Federal Nuclear Center–VNIIEF, Sarov,
Nizhniĭ Novgorod Region
Moscow Institute of Physics & Technology, Dolgoprudnyĭ,
Moscow Region, Russia

Abstract

This essay is written on the occasion of the 160th anniversary of the birth of Hermann Minkowski and devoted to the analysis of his creative heritage from the perspective of modern physics. Considerable attention is paid to the issue of relevance and optimal methods for teaching the four-dimensional picture of event space in high schools and universities.

1. Historical Introduction

This year marks the 160th anniversary of the birth of Hermann Minkowski, known among physicists mainly as the author of a geometric treatment of the theory of relativity with the aid of a four-dimensional continuum, the universal image of

*Corresponding Author's Email: nalewajs@chemia.uj.edu.pl

In: Mathematical Problems in Relativity, Gravitation, and Cosmology
Editor: Valeriy Dvoeglazov
ISBN: 979-8-89530-622-2

space and time. We call this continuum *Minkowski space* and denote it by $\mathbb{R}_{1,3}$ [1]. Minkowski himself used the name *world* for $\mathbb{R}_{1,3}$, and this term took root in a number of derived concepts, such as world line, world surface, and world volume. In his talk at the 80th Assembly of German Natural Scientists and Physicians on 21 September 1908 Minkowski asserted: 'Space by itself, and time by itself, are doomed to fade away into mere shadows, and only a kind of union of the two will preserve an independent reality' [1]. Minkowski expected that physicists will become imbued with a four-dimensional way of thinking, and this will lead to a revolutionary change in the understanding of physical reality. However, physicists were in no hurry to do this. Suffice it to recall that Albert Einstein did without $\mathbb{R}_{1,3}$ until 1912, considering this geometric construction as a manifestation of 'excessive learning' (uberflüssige Gelehrsamkeit) [2], and took the four-dimensional picture in earnest only after he identified gravity with the curvature of spacetime [3].

Admittedly, even today recognition of this great discovery can be found among just professional researchers, whereas the vision of the geometry of physical reality from the scientific position of the late 19th century predominates in university curricula.

The idea of fusing space and time was not so new to mathematicians and physicists who were contemporaries of Minkowski. Felix Klein used a four-dimensional description as a convenient mathematical device in his lectures on the theory of the top, where time played the role of the fourth dimension [4]. Furthermore, in 1764 Jean Leron d'Alembert noted the possibility of considering time as a fourth dimension in his article 'Dimensionality' for the *Encyclopedia of Sciences, Arts and Crafts*.

At first glance, the concepts of space and time are in no way connected, and at best their unity is purely verbal. However, upon careful examination a connection can be discovered. The connecting link turns out to be the concept of *movement*. Ancient thinkers guessed this. Zeno's famous aporias can serve clear evidence of their insights. To reconcile the real and logical pictures of infinitely slow motion, a more penetrating insight into of the properties of space and time was required.

The turning point occured at the beginning of the 20th century. By then the idea of an *upper limit* of speed was ripe. The maximum speed of movement was attributed to the speed of light in a vacuum, c. Bodies can move with a speed v not exceeding c. No signal can overtake a light signal. In general, any local energy-momentum impact is transmitted in space no faster than at a rate c. Thus, the unity of space and time was revealed not only in the infinitely slow movement associated with Zeno's aporia, but also in the need to spend finite time to move in space. In his lecture to a Congress of Arts and Sciences at St. Louis on 24 September 1904 Henri Poincaré [2]

[1] Another designation, $\mathbb{M}_4$, introduced in honor of Minkowski, is also popular in the physics literature.
[2] Note that this year marks the 170th anniversary of the birth of Poincaré.

predicted an era of ‘completely new dynamics, which, among other things, will be characterized by the rule that no speed can exceed the speed of light’ [5].

Could the idea of a maximum speed have come about earlier? For its appearance it would not even be necessary to invent methods for measuring speeds. This is visible to the naked eye as in an ancient allegory: Achilles is faster than a tortoise, a dart is faster than Achilles, etc., and a thunderbolt is faster than any of them. However, a flash of lightning will not catch up with another flash of lightning released earlier. Mathematically, we have a chain $t < A < d < \cdots < l$ whose elements are connected by the order relation ‘$<$’. The fact that $l < l$ means that there is a supremum, l, of this set.

The idea of a maximum speed c was contrary to the *principle of relativity* formulated by Galileo. According to this principle, when passing from one inertial frame of reference L to another one L', the velocity $\mathbf{v}$ of a given object must be added to the relative velocity $\mathbf{V}$ of these frames, $\mathbf{v}' = \mathbf{v} + \mathbf{V}$. This implies that the front of a given wave of light moves at different rates in L and L'. The task, therefore, was to reconcile the principle of relativity with the treatment of c as an universal constant.

Einstein [6] and Poincaré [7] successfully completed the task, creating the special theory of relativity. Following Minkowski, the essence of this theory is expressed in one phrase:

> *Space and time form a single four-dimensional continuum, the event space $\mathbb{R}_{1,3}$, described by pseudo-Euclidean geometry.*

This article is an expression of the deepest respect for the creator of $\mathbb{R}_{1,3}$, the concept that has left an indelible imprint on modern physics. The article summarizes the material of the author’s works published in journals of various ranks and addressed to different audiences. It involves issues that may seem ‘low-level’ to some readers with issues that may be troublesome for other readers. And yet the author cherishes the humble hope that reading this rather long article will benefit those who have the patience to reach the end.

2. Advantages of $\mathbb{R}_{1,3}$

Friedrich Wilhelm Leibniz regarded our world as ‘the most perfect of all possible worlds’ (Le meilleur des mondes possibles) [8]. Nowadays this guess about God’s providence could turn into an immutable intellectual conviction that $\mathbb{R}_{1,3}$ is the most perfect mathematical *model* of possible worlds. However, we put aside the question of supreme harmony for a while, and discuss the pedagogical value of $\mathbb{R}_{1,3}$ using examples from the university course on theoretical physics [9]–[12].

2.1. Brevity is the Soul of Wit

Physics students get excited when they first see how simple Maxwell's equations look in four-dimensional tensor form [3],

$$\partial_\mu F^{\mu\nu} = 4\pi j^\nu\,, \tag{1}$$

$$\partial_\mu {}^*F^{\mu\nu} = 0\,, \tag{2}$$

as compared to their traditional three-dimensional vector notation,

$$\operatorname{div}\mathbf{E} = 4\pi\varrho\,, \quad \operatorname{curl}\mathbf{B} = 4\pi\mathbf{j} + \frac{\partial\mathbf{E}}{\partial t}\,, \tag{3}$$

$$\operatorname{div}\mathbf{B} = 0\,, \quad \operatorname{curl}\mathbf{E} = -\frac{\partial\mathbf{B}}{\partial t}\,, \tag{4}$$

and are just fascinated by their record with the aid of Cartan differential forms,

$$d\,{}^\star F = 4\pi J\,, \tag{5}$$

$$dF = 0\,, \tag{6}$$

where the 2-forms $F = \frac{1}{2}\,F_{\mu\nu}\,dx^\mu \wedge dx^\nu$ and ${}^\star F = \frac{1}{2}\,{}^*F_{\mu\nu}\,dx^\mu \wedge dx^\nu$, together with the 3-form $J = \frac{1}{6}\,J_{\lambda\mu\nu}\,dx^\lambda \wedge dx^\mu \wedge dx^\nu$ related to j^μ by the relation $J_{\lambda\mu\nu} = \epsilon_{\lambda\mu\nu\rho}\,j^\rho$, are used.

To moderate this enthusiasm, one usually offers the following argument. The dynamical law governing the electromagnetic field is encoded in Eqs. (1) and (2) more compactly than it is in Eqs. (3) and (4). It is like that. But which of these equations is more convenient for their decoding, that is, for finding their solutions?

Surprisingly, it is easier to outline a strategy for finding solutions to Eqs. (5) and (6) than to Eqs. (3) and (4). Formally, we are dealing with a seemingly *overdetermined* set of partial differential equations: 8 equations are used to find 6 unknown functions (6 coordinates of vectors $\mathbf{E}$ and $\mathbf{B}$ or, equivalently, 6 components of the antisymmetric tensor $F^{\mu\nu}$). An important step is to convert it to a *determined* set of equations. The identity $dd \equiv 0$ for the differential 1-form d contains a hint that the 2-form

$$F = dA\,, \tag{7}$$

[3] We accept the metric $\eta_{\mu\nu} = \operatorname{diag}(+1, -1, -1, -1)$. Gaussian units are used throughout. The speed of light c and Planck's constant $\hbar$ are set equal to 1. The tensor ${}^*F^{\lambda\mu}$, dual to the electromagnetic field strength, is defined by ${}^*F^{\lambda\mu} = \frac{1}{2}\,\epsilon^{\lambda\mu\nu\rho}\,F_{\nu\rho}$. Einstein's rule that a repeated index is summed over applies. Most of the notations are standard.

where $A = A_\mu dx^\mu$ is an arbitrary smooth 1-form, gives a general solution to Eq. (6). The four-vector $A_\mu = A_\mu(x)$ is called the *vector potential* of the electromagnetic field. In coordinate notation, Eq. (7) takes the form:

$$F_{\rho\sigma} = \partial_\rho A_\sigma - \partial_\sigma A_\rho \,. \tag{8}$$

This construction renders Eq. (2) an identity because the contraction of a symmetric tensor $\partial_\mu \partial_\rho$ with the antisymmetric tensor $\epsilon^{\mu\nu\rho\sigma}$ vanishes identically, $\epsilon^{\mu\nu\rho\sigma}\partial_\mu\partial_\rho \equiv 0$. In this regard, Eq. (2) is often called the *Bianchi identity*.

The pertinent ansatz that turns Eqs. (4) into an identity is not so obvious. True, if we write the components of A^μ in a fixed inertial frame, $A^\mu = (\phi, \mathbf{A})$, then taking into account the definitions $E_i = F_{0i}$ and $B_i = -\frac{1}{2}\,\epsilon_{ijk}F^{jk}$ we can express Eq. (8) in three-dimensional vector form,

$$\mathbf{E} = -\frac{\partial \mathbf{A}}{\partial t} - \nabla\phi \,, \quad \mathbf{B} = \mathrm{curl}\,\mathbf{A} \,. \tag{9}$$

By substituting (9) into (4), we verify that this equation is identically satisfied for any smooth ϕ and $\mathbf{A}$. However, to invent the ansatz (9) as a general solution to Eq. (4) considerable mathematical sophistication is required.

Equation (8) does not uniquely define A_μ. Indeed, replacing A_μ with a new vector function A'_μ,

$$A_\mu \to A'_\mu = A_\mu + \partial_\mu \chi \,, \tag{10}$$

where $\chi(x)$ is an arbitrary smooth function, leaves the observed quantity $F_{\lambda\mu}$ unchanged. The replacements of A_μ with A'_μ of the form of Eq. (10) are called *gauge transformations*. Therefore, Eq. (8) introduces the whole equivalence class of functions related by the gauge transformations (10), rather than a certain function. The terms $\partial_\mu\chi$ in Eq. (10) are called *gauge modes*.

Substituting (8) into (1) gives

$$\Box A^\mu - \partial^\mu \partial^\nu A_\nu = 4\pi j^\mu \,, \tag{11}$$

where $\Box \equiv \partial^\lambda \partial_\lambda = \partial^2/\partial t^2 - \nabla^2$ is the wave operator. We are thus led to the system of equations (8) and (11), in which the number of equations is equal to the number of unknown functions. This amendment of the system of equations results from augmenting the field degrees of freedom by gauge variables which are *auxiliary* degrees of freedom. These variables do not affect the dynamics of particles because they do not contribute to the Lorentz force $qF^{\mu\nu}v_\nu$. Moreover, the current j^ν is not a source of gauge modes. The evolution of χ is unrelated to that of $F^{\mu\nu}$. We are thus entitled to choose gauge modes to suit our convenience, for example, from all elements of a given equivalence class of A_μ, we can take a single representative.

Let us rewrite Eq. (11) as

$$\Pi^{\mu\nu}A_\nu \equiv (\Box\,\eta^{\mu\nu} - \partial^\mu\partial^\nu)\,A_\nu = 4\pi j^\mu\,. \tag{12}$$

If there were an operator Π^{-1}, the inverse of Π, then the solution to Eq. (12) would have the form $A = 4\pi\Pi^{-1}j$. However, $\det\Pi = 0$, that is, the operator Π does not have an inverse, and so it is impossible to solve Eq. (12) directly.

How to find solutions to this equation? Let us take advantage of the freedom in choosing gauge modes. We can impose an additional condition on A^μ to narrow the limits of gauge arbitrariness, for example, to use the Lorenz gauge-fixing condition

$$\partial_\nu A^\nu = 0\,. \tag{13}$$

Then Eq. (11) becomes an inhomogeneous *wave equation*,

$$\Box A^\mu = 4\pi j^\mu\,. \tag{14}$$

Regular methods for solving this equation are discussed extensively in textbooks of mathematical physics, so we will not dwell on them here.

This offers a clearer view of how the student could become familiar with the idea of *gauge fields* by the example of A_μ. This acquaintance in the course of unveiling the laws of electrodynamics is the first step in an organic immersion into the geometric description of the fundamental forces of nature.

However, in a number of textbooks on theoretical physics, the electromagnetic field is defined, from the outset, in terms of vector potentials A_μ [4]. This approach seems of doubtful value in teaching students who have mastered the course of general physics, say, by careful study of the textbooks [14], and are eager for the logical ordering of this baggage in the framework of theoretical physics. The mathematician defines A_μ as a linear connection in fiber spaces with the Abelian structure group $\mathrm{U}(1)$, that is, the quantity that is by no means so simple to be accepted as a primary concept. Perhaps someone will object that there is no other way to define the electromagnetic field. In the context of this disagreement on the basic tenets of classical electrodynamics it seems wise to examine this subject more thoroughly, step by step.

2.2. Kinematics and Dynamics

A primary object in Minkowski world is a point particle. Its history is depicted by a world line. The world line of a massive particle is a smooth timelike curve in

[4] This is the case, in particular in an excellent textbook by Lev Davidovich Landau and Evgeniĭ Mikhaĭlovich Lifshitz [13], where the familiarity of the reader with the electromagnetic field in §16 begins with the phrase: 'The properties of the field are characterized by a 4-*vector* A_i, the so-called 4-potential'.

$\mathbb{R}_{1,3}$, and world lines of massless particles [5] can be associated with smooth lightlike curves [15]. The world line is a much simpler mathematical object than its associated three-dimensional trajectory. One can conceive the trajectory of a fly that entered into a room through a window, flew around the room many times along a tangled path, and left the room through the same window. If we pull the two ends of this trajectory in opposite directions, we will get a straight line with k nodes on it, with k being arbitrary. With a similar stretching of the corresponding world line, we obtain a straight line with no nodes. A simple reason for this is that a smooth timelike curve never changes its orientation relative to the time axis, and hence, cannot form loops which are potentially suitable for the occurrence of tied knots[6]. Therefore, the topological properties of the world line differ noticeably from those of its associated three-dimensional trajectory. We will not deplete the content of classical field theory if we assume that every world line is an infinitely differentiable curve. A trajectory is not such a simple object in the differential-geometric respect because cusped trajectories cannot be excluded from consideration in advance. For example, the trajectory of a stone moving up and down along a vertical direction involves a cusp. This line may result from a sequence of smooth trajectories of the stone thrown at an angle α to the horizon as $\alpha \to \frac{\pi}{2}$. The corresponding family of world lines, including the limiting case, is described by smooth curves.

We now turn to the dynamics. The equation of motion for a relativistic particle

$$\frac{d}{dt}\left(\frac{m\mathbf{v}}{\sqrt{1-\mathbf{v}^2}}\right) = \mathbf{F} \tag{15}$$

is due to Poincaré [7] and Max Planck [17]. Minkowski introduced a more convenient parameter of evolution, the *proper* time s, related to laboratory time t by

$$ds = \sqrt{1-\mathbf{v}^2}\, dt\,, \tag{16}$$

and the concept of *four-velocity*, $v^\mu = dz^\mu/ds$, where $x^\mu = z^\mu(s)$ is the parametric equation of the world line. This parametrization imposes the constraint

$$v^2 = 1\,, \tag{17}$$

which stems from $dz^2 = ds^2$. Geometrically, Eq. (17) shows that the history of a particle is traced by a curve on the upper part of the four-dimensional hyperboloid $v_0^2 - \mathbf{v}^2 = 1$.

---

[5]It is unlikely that in the days of Minkowski, anyone would have dared to talk about interacting massless particles. Presently this is no longer a mind game but rather completely meaningful subject for research. In the quark-gluon plasma, chiral symmetry is restored, and this fact is likely to evidence that massive quarks convert into their massless incarnations.

[6]Note also that any curve in Euclidean space $\mathbb{R}_n$ is free of knots when $n \geq 4$ [16].

But, what more important, we owe the idea that Eq. (15) is nothing new as compared to Newton's second law

$$\frac{d\mathbf{p}}{dt} = \mathbf{f} \tag{18}$$

to Minkowski. To derive Eq. (15) it is unnecessary to abandon Eq. (18) and even modify it. It is sufficient to *embed* Eq. (18) smoothly into $\mathbb{R}_{1,3}$ [7].

The idea of embedding comes from the assumption that Eq. (18) is an exact law in the limit $\mathbf{v} \to 0$. If we fix an *instantaneously comoving* reference frame, then we can precisely predict the evolution of the particle in this frame during an ensuing evanescent time interval. In geometric language, this means that Eq. (18) is a strict vector relation on a hyperplane Σ perpendicular to the world line of the particle. But Σ tilts together with its normal vector v^μ as one moves along the curve. To recover the global evolution one joins together the small fragments from each of the instantaneously comoving frames. The algorithm for constructing a world line is as follows: at the initial moment $s = s_0$, the rest frame of the particle is fixed, a fragment of the curve is found using Eq. (18), then the rest frame is fixed at a close moment $s = s_0 + \epsilon$, a new fragment of the curve is calculated, etc. This is reminiscent of a movie, which is actually a discrete set of pictures representing the dynamical process at the hyperplanes Σ.

For the embedding to be made smooth, we need an operator $\overset{v}{\perp}$ which incessantly projects four-vectors on hyperplanes Σ perpendicular to the world line. The operator is

$$\overset{v}{\perp}_{\mu\nu} = \eta_{\mu\nu} - \frac{v_\mu v_\nu}{v^2}\,. \tag{19}$$

Note that $\overset{v}{\perp}$ is the same for any parametrization. If v_μ is replaced by $\dot{z}_\mu = dz_\mu/d\tau$, where τ is another parameter of evolution, this leaves the form of $\overset{v}{\perp}$ unchanged.

The time axis in the instantaneously comoving frame is parallel to the tangent of the world line, hence dt equals ds. Accordingly, in the instantaneously comoving frame, the derivative with respect to t may be replaced by the derivative with respect to s.

Given the three-dimensional vector $\mathbf{f}$ in the hyperplane Σ, one can unambiguously construct a four-dimensional vector f^μ. Indeed, in the instantaneously comoving frame,

$$f^\mu = (0,\, \mathbf{f})\,. \tag{20}$$

---

[7] In this respect, some textbooks mislead the novice. For example, in Ref. [18], Chapter 5, one can read: 'The laws of Newtonian mechanics have to be changed to be consistent with the principles of special relativity' and 'The objective of relativistic mechanics is to introduce the analogue of Newton's second law $\vec{F} = m\vec{a}$. There is nothing from which this law can be *derived*, but plausibly it must satisfy certain properties...' These statements are erroneous.

In an arbitrary inertial frame, components of f^μ can be found from (20) through the appropriate Lorentz boost. f^μ is called the *Minkowski force* or *four-force*.

In the rest frame, $v^\mu = (1, \mathbf{0})$. Combining this with Eq. (20), we find

$$v \cdot f = 0\,. \tag{21}$$

Since $v \cdot f$ is an invariant, the Minkowski force is orthogonal to the four-velocity in any Lorentz frame.

Let us define the *four-momentum* of the particle p^μ. Our concern here is, in fact, with the derivative of the four-momentum with respect to the proper time, dp^μ/ds. We require that its spatial components in the instantaneously comoving frame dp^i/ds be identical to the respective components of $d\mathbf{p}/dt$ in Eq. (18). Note that dp^0/ds remains indeterminate in this frame.

The embedding of Newton's second law (18) in hyperplanes perpendicular to the world line is given by

$$\overset{v}{\perp}_{\mu\nu}\left(\frac{dp^\nu}{ds} - f^\nu\right) = 0\,. \tag{22}$$

This is the desired *dynamical law* for a relativistic particle. In symbolic form,

$$\overset{v}{\perp}(\dot{p} - f) = 0\,. \tag{23}$$

The presence of $\overset{v}{\perp}$ in Eq. (23) suggests that we have three independent equations. Indeed, when contracted with v^μ, the four equations reveal a linear relation resulting from

$$v^\mu\left(\eta_{\mu\nu} - \frac{v_\mu v_\nu}{v^2}\right) \equiv 0\,.$$

Isaac Newton postulated that $\mathbf{p}$ is proportional to $\mathbf{v}$,

$$\mathbf{p} = m\mathbf{v}\,. \tag{24}$$

The coefficient of proportionality m measures the inertia of the particle. Objects with the momentum of this type will be called *Galilean particles* or simply *particles*. In the four-dimensional picture, this notion is slightly reformulated. We refer to the Galilean particle as a point object that possesses the four-momentum [8]

$$p^\mu = mv^\mu\,. \tag{25}$$

[8]Classical (not quantum) theory allows for the existence of non-Galilean objects. Such objects have four-momenta p^μ whose dependence on kinematic variables is different from that shown in Eq. (25). Readers interested in these objects may consult [12] and [9].

Taking into account the identity $v \cdot a = 0$, which stems from differentiation of Eq. (17), we conclude that if p^μ is given by Eq. (25), then the projector $\overset{v}{\perp}$ in Eq. (22) acts as the unit operator, so that this equation becomes

$$ma^\mu = f^\mu \,. \tag{26}$$

It is convenient to separate the Lorentz factor $\gamma = (1 - \mathbf{v}^2)^{-1/2}$ as an overall factor in f^μ,

$$f^\mu = \gamma\,(\mathbf{F}\cdot\mathbf{v},\,\mathbf{F})\,. \tag{27}$$

With $ds = \gamma^{-1}dt$, we find that the spatial component of Eq. (26),

$$\frac{d}{dt}\,(m\gamma\mathbf{v}) = \mathbf{F}\,, \tag{28}$$

is identical to the Poincaré–Planck equation (15). The time component of Eq. (26),

$$\frac{d}{dt}\,(m\gamma) = \mathbf{F}\cdot\mathbf{v}\,, \tag{29}$$

might be interpreted as the variation of the particle energy $\varepsilon = m\gamma$ due to the work performed by the force $\mathbf{F}$ in a unit time. A direct calculation shows that Eq. (29) follows immediately from Eq. (28).

2.3. How to Define the Electromagnetic Field?

The student embarking on a study of electrodynamics in a course of theoretical physics may be surprised that the textbooks do not begin with a definition of the electromagnetic field. Strange though it may seem, this definition is also missing from the later stages of the physics curriculum. For example, a graduate course in electromagnetism covered by John David Jackson's book [19], which is informally accepted as *the* standard textbook of classical electrodynamics in Western universities, lacks a well defined concept of the electromagnetic field. To introduce the electric field strength $\mathbf{E}$ and magnetic induction $\mathbf{B}$, Jackson [9] refers to the Lorentz force equation

$$\mathbf{F} = q\,(\mathbf{E} + \mathbf{v}\times\mathbf{B})\,, \tag{30}$$

and notes that this formula was originally discovered experimentally and then confirmed in numerous experimental studies. The impression gained from such an introduction of $\mathbf{E}$ and $\mathbf{B}$ is that the defining equation (30) is well substantiated phenomenologically but has no theoretical motivation. Why do we need a theoretical motivation? A phenomenological finding may be merely an approximation

[9] To be specific, we mean the second edition of this book, pp. 3-4.

for a fundamental relationship. If one day we had to correct the analytic form of Eq. (30), then the very sense of what we held to be the electromagnetic field may drastically change along with it. An unambiguously formulated and theoretically motivated definition is really necessary because otherwise we discuss the properties of a physical object while being ignorant of its essence, and hence we run the risk of confusing it with objects of a different nature.

It is impossible to define the electromagnetic field without relying on theoretical power of the quantities v^μ, f^μ, and $F_{\mu\nu}$, introduced by Minkowski. Since the concept of a particle is primary, the definition should display how the field exerts on particles. The key observation is that f^μ is orthogonal to v^μ at the point of application of the four-force. Therefore, f^μ *cannot but depend* on v^μ. The simplest forms of the dependence of f^μ on v^μ are given by linear and quadratic functions. Let f^μ be linear in v^μ. The case that $f^\mu = \alpha v^\mu$ is not of interest because f^μ is orthogonal to v^μ only for $\alpha = 0$. We then turn to $f^\mu = \beta^{\mu\nu} v_\nu$. From Eq. (21) we find that

$$\beta^{\mu\nu} v_\mu v_\nu = 0 \,. \tag{31}$$

This equation is satisfied for any v^μ provided that

$$\beta^{\mu\nu} = -\beta^{\nu\mu} \,. \tag{32}$$

We thus see that if a physical object is distributed throughout spacetime and acts on a particle through a four-force f^μ linear in the four-velocity v^μ, then this object is characterized by an antisymmetric tensor $\beta^{\mu\nu}$ at each point of $\mathbb{R}_{1,3}$. Such objects are collectively known as force fields, or simply fields.

However, $\beta^{\mu\nu}$ contains information on both the state of the field and how it affects the particle. These quantities should be separated. In the simplest case, the coupling of a particle with the field is given by a *scalar* parameter q,

$$f^\mu = q F^{\mu\nu} v_\nu \,. \tag{33}$$

This possibility is actually realized in nature. The field whose state at each point x^μ of $\mathbb{R}_{1,3}$ is specified by an antisymmetric tensor $F_{\mu\nu}$ and its action on a particle is given by the four-force f^μ of the form of Eq. (33) is called *electromagnetic* field, and $F_{\mu\nu}$ is referred to as the electromagnetic field strength. The scalar quantity q will be provisionally called the electric *charge-coupling*.

We will assume that the particle remains identical to itself throughout its entire history. This implies, in particular that the coupling of the particle and the field does not change over time,

$$\dot{q} = 0 \,. \tag{34}$$

In a particular inertial frame of reference, two three-dimensional vectors, $\mathbf{E}$ and $\mathbf{B}$, can be always defined in terms of six components of the antisymmetric tensor $F_{\mu\nu}$,

$$E_i = F_{0i} = F^{i0}\,, \quad B_k = -\frac{1}{2}\,\epsilon_{klm}F^{lm}\,. \tag{35}$$

where ϵ_{klm} is the three-dimensional Levi-Civita symbol. From (33) we can obtain (30). Both quantities f^μ and $\mathbf{F}$, expressed respectively by Eqs. (33) and (30), go under the general name of Lorentz force.

The electromagnetic field is thus defined as a field of the *simplest form* in $\mathbb{R}_{1,3}$ [10]. One may wonder why simplicity is elevated to the rank of a fundamental principle. To answer this question, we recall that many fundamental physical principles are statements about extreme values of some physical quantities, as exemplified by the principle of least action in mechanics or the second law of thermodynamics selecting the state of maximal entropy. Simplicity is a kind of extremal quality, and so the search for the simplest form of f^μ is fully justified.

On the other hand, if we would look for the simplest form of the three-force $\mathbf{F}$ defined in a particular frame of reference by the conventional decomposition $f^\mu = \gamma\,(\mathbf{F}\cdot\mathbf{v},\mathbf{F})$, then this attempt would be foiled. Indeed, $\mathbf{F}$ need not be velocity-dependent, so that the simplest form of this quantity is $\mathbf{F}(\mathbf{x}) = \mathbf{F}_0$. However, Eq. (30) defies all attempts at elevating it to a satisfactory definition of the electromagnetic field. The concept of the electromagnetic field, naturally arising in $\mathbb{R}_{1,3}$, seems to be devoid of theoretical motivations beyond this context.

A slightly more involved case is that the coupling of a particle and the electromagnetic field is characterized by a *pseudoscalar* parameter $q^\star$. Consider a four-force of the form

$$f^\mu = q^\star v_\nu\, {}^*F^{\mu\nu}\,, \tag{36}$$

where ${}^*F^{\lambda\mu} = \frac{1}{2}\,\epsilon^{\lambda\mu\nu\rho}F_{\nu\rho}$. We adopt the rule that the three-dimensional and four-dimensional Levi-Civita symbols are related by

$$\epsilon_{ijk} = \epsilon^{0ijk}\,. \tag{37}$$

Taking into account Eq. (35), we find

$$\mathbf{F} = q^\star\,(\mathbf{B} - \mathbf{v}\times\mathbf{E})\,. \tag{38}$$

The parameter $q^\star$ behaves as a pseudoscalar under space reflections. This parameter is called magnetic charge-coupling, and a particle experiencing the action of a three-force $\mathbf{F}$ defined by Eq. (38) is called *magnetic monopole.*

[10] Some clarification is necessary here. By the simplest field is meant an agent responsible for interaction such that oppositely charged particles attract, and charges of the same signs repel. In the next subsection we will discuss a different picture in which charges of the same sign attract. Then a scalar field turns out to be the simplest carrier of the interaction.

Note that Eq. (36) is not to define a new field; an ordinary electromagnetic field appears here. This formula gives a definition of a particle carrying a magnetic charge $q^\star$.

Much effort has been spent on the experimental search for magnetic monopoles, but no evidence of their existence has been found. Therefore, at present such particles have the status of hypothetical objects.

And yet, the acquaintance of students with magnetic monopoles using Eq. (36) is not a waste of time. Modern physics not only studies the properties of observed particles, but also tries to gain insight into why the closest relative of a charged particle, the magnetic monopole, is denied a residence permit in $\mathbb{R}_{1,3}$.

2.4. The Other Fundamental Interactions

We next look at a particle possessing additional degrees of freedom, which is connected with a field through a *vector coupling* Q_a. Let V be a vector space of dimension $\mathcal{N}$, the so-called *internal space*. For example, in the *weak interactions*, V is a three-dimensional space that is called *isospin* space, and in the *strong interactions*, V has eight dimensions and is called *color* space. The most general form of f^μ linear in v^μ can be written [20] as

$$f_\mu = \sum_{a=1}^{\mathcal{N}} Q_a \, v^\nu \, G^a_{\mu\nu} \,, \tag{39}$$

where $G^a_{\mu\nu}$ specifies states of the field and is called the *gauge field* strength or *Yang–Mills field* in honor of its discoverers, Chen-Ning Yang and Robert Mills [21].

However, the vector coupling in Eq. (39) is too arbitrary to generate a meaningful dynamics. Therefore, an additional constraint is to be imposed on V, namely V is endowed with the structure of a *Lie algebra*. To put it simply, elements of V may not only be added but also multiplied among themselves. Let $\mathcal{N}$ elements $\{T_a\}$ span a basis of the vector space V. Any vector A belonging to V can be expanded over this basis, $A = A^a T_a$. The following multiplication rule for the basis elements holds:

$$[T_a, T_b] = i \sum_a f^c_{\ ab} \, T_c \,, \tag{40}$$

where $f^c_{\ ab} = -f^c_{\ ba}$ are numerical coefficients called *structure constants* of the Lie algebra. Thus, $[T_a, T_b]$ is an anticommutative bilinear operation. This operation is not associative, that is, $[T_a, [T_b, T_c]]$ is not equal to $[[T_a, T_b], T_c]$. In lieu of the associativity in the theory of Lie algebras, the Jacobi identity is adopted,

$$[T_a, [T_b, T_c]] + [T_c, [T_a, T_b]] + [T_b, [T_c, T_a]] = 0 \,. \tag{41}$$

If $f^c_{\ ab} = 0$, as exemplified by an algebra with a basis containing a single element, then the Lie algebra is called *Abelian*. Such is the case for the electromagnetic interactions. If not all structure constants vanish, then the Lie algebra is called *non-Abelian*. Accordingly, $G_{\mu\nu} = G^a_{\mu\nu} T_a$ is referred to as a *non-Abelian field*. It is expressed in terms of *non-Abelian vector potentials* $A_\mu = A^a_\mu T_a$,

$$G_{\mu\nu} = \partial_\mu A_\nu - \partial_\nu A_\mu + ig\,[A_\mu, A_\nu]\,, \tag{42}$$

where g is the so-called Yang–Mills coupling constant. In this regard, it is customary to say that the *vector* fields A_μ mediate gauge interactions. The electromagnetic, weak, and strong interactions are carried by vector fields.

To complete our discussion, we should define the *gravitational* field. With this aim in mind the four-force f^μ is to be quadratic in the four-velocity v^μ. The necessity of considering quadratic relationships is due to the empirical fact that identical gravitating particles attract each other. With reference to Exercise 7.2 in [22], it transpires that if the interaction is carried by $F_{\mu\nu}$ (or $G^a_{\mu\nu}$), then identical particles with a real coupling q (or Q^a) repel each other. In contrast, quadratic relationships between f^μ and v^μ are compatible with scalar and symmetric second rank tensor fields mediating such interactions that ensure the attraction of identical particles (see Exercises 7.1 and 7.3 in [22]).

Consider a particle interacting with a scalar field $\phi(x)$. We wish to extend the notion of the potential force $\mathbf{f} = -\nabla\Phi$, commonly used in Newtonian mechanics, to the relativistic context. Let f^μ be orthogonal to v^μ and contain gradients of the scalar field,

$$f_\mu = m_{\rm g}\left(\partial_\mu\phi - v_\mu v^\nu \partial_\nu\phi\right), \tag{43}$$

where $m_{\rm g}$ is a charge-coupling linking the particle with the field ϕ. If the particle is at rest, $v^\mu = (1, \mathbf{0})$, then Eq. (43) becomes $f^\mu = (0, -m_{\rm g}\nabla\phi)$. Therefore, a four-dimensional analogue of the Newtonian potential force is the four-force f^μ which is quadratic in the four-velocity v^μ. The second term in the parenthesis of Eq. (43) can be converted to

$$-v_\mu v^\nu \partial_\nu\phi = -v_\mu \frac{dz^\nu}{ds}\frac{\partial\phi}{\partial z^\nu} = -v_\mu \frac{d\phi}{ds}\,. \tag{44}$$

If we add to f_μ defined by (43) the term ϕa_μ consistent with the orthogonality conditions (21), then the equation of motion for a particle interacting with a scalar field becomes

$$\frac{d}{ds}\left(m + m_{\rm g}\phi\right) v^\mu = m_{\rm g}\partial^\mu\phi\,. \tag{45}$$

The fact that the four-force is quadratic in the four-velocity is here implicit:

Our concern here is with the parameter $m_{\rm g}$ consistent with the Einstein equivalence principle which maintains that $m_{\rm g}$ is proportional to the inertial mass of the particle,

$$m_{\rm g} = km\,, \tag{46}$$

where k is a numerical coefficient whose value depends on the choice of units. We redefine the scalar field, $\phi \to \phi' = 1 + k\phi$, to rearrange Eq. (45) to

$$\frac{d}{ds}\left(\phi v^{\mu}\right) = \partial^{\mu}\phi\,. \tag{47}$$

This equation governs a particle in Gunnar Nordström's theory of gravity [23] where the scalar field ϕ is responsible for the gravitational interaction. This is the *simplest* theory of gravity. Unfortunately, this theory is unable to correctly take into account some observed gravitational effects, such as deflection of a light ray as it passes near the Sun.

We now turn to a generic case that the four-force is quadratic in the four-velocity,

$$f^{\lambda} = -m_{\rm g}\Gamma^{\lambda}_{\mu\nu}v^{\mu}v^{\nu}\,, \tag{48}$$

where $\Gamma^{\lambda}_{\mu\nu}$ is a tensor symmetric in μ and ν. By Eq. (21),

$$\Gamma_{\lambda\mu\nu}v^{\lambda}v^{\mu}v^{\nu} = 0\,. \tag{49}$$

For this equation to hold for any v^{μ}, it is necessary that the sign of $\Gamma_{\lambda\mu\nu}$ must change under permutation of λ and μ and that of λ and ν. However, any tensor $\Gamma_{\lambda\mu\nu}$ which is symmetric in μ and ν, and antisymmetric in λ and μ as well as in λ and ν, is zero. Indeed, any cyclic permutation of indices changes the sign of this tensor, $\Gamma_{\lambda\mu\nu} = -\Gamma_{\nu\lambda\mu}$. Having performed three cyclic permutations of the indices, we arrive at the original location of the indices. On the other hand, this operation changes the sign of $\Gamma_{\lambda\mu\nu}$ three times, so that $\Gamma_{\lambda\mu\nu} = -\Gamma_{\lambda\mu\nu}$, which implies that $\Gamma_{\lambda\mu\nu} = 0$.

Is it possible to improve the line of reasoning in such a way that proceeding from Eq. (48) we would obtain $\Gamma_{\lambda\mu\nu}$ which is not zero? Such is actually the case. Expression (48) may be nontrivial if a pseudo-Riemannian metric $g_{\mu\nu}(x)$ of a curved spacetime manifold, rather than the pseudoeuclidean metric $\eta_{\mu\nu}$ is concerned. Now the line element is

$$ds^2 = g_{\mu\nu}(x)dx^{\mu}dx^{\nu}\,. \tag{50}$$

Let the world line be timelike. Then the condition that a vector tangent to this curve $v^{\mu} = dz^{\mu}/ds$ has unit length is

$$g_{\mu\nu}v^{\mu}v^{\nu} = 1\,. \tag{51}$$

It follows that

$$\frac{d}{ds}\left(g_{\mu\nu}v^{\mu}v^{\nu}\right)=v^{\mu}v^{\nu}\,\frac{d}{ds}\,g_{\mu\nu}+2g_{\mu\nu}v^{\mu}a^{\nu}=v^{\mu}v^{\nu}v^{\lambda}\partial_{\lambda}g_{\mu\nu}+2g_{\mu\nu}v^{\mu}a^{\nu}=0\,. \tag{52}$$

Let us introduce the *Christoffel symbol*

$$\Gamma_{\lambda\mu\nu}=\frac{1}{2}\left(\partial_{\mu}g_{\nu\lambda}+\partial_{\nu}g_{\lambda\mu}-\partial_{\lambda}g_{\mu\nu}\right), \tag{53}$$

substitute (46) into (48), and put $k=1$. Then the equation of motion for a particle affected by $\Gamma_{\lambda\mu\nu}$,

$$\frac{dv_{\lambda}}{ds}+\Gamma_{\lambda\mu\nu}v^{\mu}v^{\nu}=0\,, \tag{54}$$

is orthogonal to the four-velocity in the sense of the metric (50). To see this, we contract Eq. (54) with v^{λ}, and use Eq. (52) to get $\frac{1}{2}\,\frac{d}{ds}\left(g_{\lambda\mu}v^{\lambda}v^{\mu}\right)=0$, as stated.

Equation (54) is the *geodetic* equation for the given metric. It governs the behavior of a test particle in the gravitational field $g_{\mu\nu}$, according to the general theory of relativity.

2.5. The Light Cone Structure

Poincaré was the first to realize the importance of quantities invariant under the group of Lorentz transformations, in particular the interval $dx^2=dx_0^2-d\mathbf{x}^2$. Minkowski was just the scientist to whom we owe the understanding of the fundamental role of the light cone. It is easy and convenient to solve the algebraic part of a problem when we are handling geometric quantities of $\mathbb{R}_{1,3}$. However, the matter is not limited to mathematical comfort. A precise definition of fundamental concepts may (on frequent occasion) be impossible outside the context of $\mathbb{R}_{1,3}$, as exemplified by the electromagnetic field. It is the light cone that offers a means for tracing the cause-and-effect relationship and the chronology of events.

Unlike Euclidean spaces which are homogeneous and isotropic, the situation in $\mathbb{R}_{1,3}$ is more complicated. We may take a point x^{μ} in $\mathbb{R}_{1,3}$ and draw a light cone $(x-y)^2=0$. Then $\mathbb{R}_{1,3}$ is divided into five regions, each enjoys the properties of homogeneity and isotropy. Those regions are: (I) the interior of the future light cone $(x-y)^2>0$, $y_0>x_0$, (II) the interior of the past light cone $(x-y)^2>0\,,\,y_0<x_0$, (III) the region outside the light cone $(x-y)^2<0$, (IV) the surface of the future light cone $(x-y)^2=0$, $y_0>x_0$, and (V) the surface of the past light cone

$(x-y)^2 = 0$, $y_0 < x_0$. This partition remains unchanged under transformations of the special orthochronous Lorentz group

$$x'^{\mu} = \Lambda^{\mu}_{\ \nu}\, x^{\nu}\,, \quad \Lambda^{\lambda}_{\ \mu}\, \eta_{\lambda\nu}\, \Lambda^{\nu}_{\ \rho} = \eta_{\mu\rho}\,, \quad \det\left(\Lambda\right) = 1\,, \quad \Lambda^{0}_{\ 0} > 0\,, \tag{55}$$

scalings and translations

$$x'^{\mu} = \lambda\, x^{\mu} + c^{\mu}\,, \quad \lambda > 0\,, \tag{56}$$

and even number of special conformal transformations

$$x'^{\mu} = \frac{x^{\mu} - b^{\mu} x^2}{1 - 2\left(b \cdot x\right) + b^2 x^2}\,. \tag{57}$$

The transformations shown in Eqs. (55)–(57) form the *conformal* group $C(1,3)$.

The converse theorem is also true [24]: if the group of generic transformations leaves the above partition of a four-dimensional normed space unchanged, then the group proves to be reducible to $C(1,3)$ with even number of special conformal transformations (57).

It is thus seen that the light cone structure is the most distinctive feature of the event space. Note that the light cone structure is more general than the metric structure $dx^2 = dx_0^2 - d\mathbf{x}^2$ of $\mathbb{R}_{1,3}$.

2.6. Simple Stories

Below we consider several issues related to $\mathbb{R}_{1,3}$, which may seem almost trivial for field theorists. As for the study of such issues at the high school and university levels, the author learned by experience that their understanding is accessible to inquisitive schoolchildren, and that mastering the four-dimensional calculation machinery should occur at the earliest stages of physics education.

2.6.1. Energy-momentum Balance in Collisions and Decays

To handle geometric quantities of $\mathbb{R}_{1,3}$, in particular four-vectors, is an enjoyable practice. We illustrate this by examining the processes presented in sections 13.3 and 13.4 of the first volume of the Berkeley course of general physics [14]. We begin with the question why a free electron cannot absorb a photon. Suppose the opposite. The notations k^{μ}, p^{μ} and p'^{μ} denote the respective four-momenta of the photon, the initial and final electron. The energy-momentum conservation as applied to photon absorption is given by

$$p^{\mu} + k^{\mu} = p'^{\mu}\,. \tag{58}$$

If k^μ is rearranged to the right side of this equation, and both sides are squared, then taking into account that $p^2 = p'^2 = m^2$, where m is the mass of the electron, and $k^2 = 0$, because the photon is a massless particle, we obtain

$$m^2 = 0 - 2k \cdot p' + m^2 \,. \tag{59}$$

The scalar product $k \cdot p'$ is most easily calculated in the electron rest frame where $p'^\mu = (m, 0, 0, 0)$, $k^\mu = (\omega, \omega\mathbf{n})$, with ω being the photon energy, and $\mathbf{n}$ a unit vector aligned with the photon direction in this frame. Therefore, $k \cdot p' = m\omega$. From (59) follows $\omega = 0$. However, a photon with $\omega = 0$ does not exist in nature. In quantum field theory, only the vacuum has zero energy. Another, more elementary argument is as follows. For the photon absorption to occur, some energy should be used to change the structure of the electron, otherwise there is no way for the energy conservation. But this is impossible for the electron which is regarded as a structureless particle.

A a natural question then arises whether the absorption of a photon by a free proton composed of three quarks, $\{u, u, d\}$, which would be excited to form a three-quark system with spin $s = \frac{3}{2}$, is feasible. The answer is positive. The anticipated three-quark system, $\Delta^+(u, u, d)$, is observed in the barion spectrum.

We next solve the following two problems. A rapidly moving proton collides with a stationary proton to create a proton-antiproton pair, so that the total number of particles in the final state is four. What is the minimum of the initial energy ε of the moving proton necessary for the reaction $\mathrm{p} + \mathrm{p} \to \mathrm{p} + \mathrm{p} + \mathrm{p} + \bar{\mathrm{p}}$ to occur? What should be the minimum of the initial energies ε, measured in the center-of-mass frame, of two protons moving towards each other in an impending head-on collision for this reaction to occur?

The key idea for the problems of minimum energy is that the particles born in these processes do not fly apart but remain nearby because their mutual separation would require additional energy which, according to the conditions of the problem, is missing from the initial particles. It is convenient to consider all finite particles in the center-of-mass frame where these particles are at rest, and their total four-momentum is given by $P^\mu = (4m, \mathbf{0})$. Accordingly, the square of this four-momentum is $P^2 = 16m^2$. The balance of four-momenta reads

$$p_1^\mu + p_2^\mu = p'^\mu_1 + p'^\mu_2 + p'^\mu_3 + p'^\mu_4 = P^\mu \,. \tag{60}$$

In a laboratory frame of reference where one of the initial protons is at rest,

$$p_1^\mu = (\varepsilon, \mathbf{p}) \,, \quad p_2^\mu = (m, \mathbf{0}) \,. \tag{61}$$

By squaring both sides of Eq. (60) we get

$$m^2 + m^2 + 2m\varepsilon = 16m^2 \quad \Longrightarrow \quad \varepsilon = 7m \,. \tag{62}$$

In the center of mass frame,

$$p_1^\mu = (\varepsilon, \mathbf{p}), \quad p_2^\mu = (\varepsilon, -\mathbf{p}), \quad P^\mu = (4m, \mathbf{0}), \tag{63}$$

and so

$$\varepsilon + \varepsilon = 4m \quad \Longrightarrow \quad \varepsilon = 2m. \tag{64}$$

Strictly speaking, processes of decays and creations of new particles are impossible in the classical picture. The reason for this is that classical dynamics is derived from the principle of least action, and the number of particles in the action is fixed. Furthermore, the photon is a particle-like excitation of the quantized electromagnetic field. The concept of photons is alien to the classical picture which treats the electromagnetic field as a smooth wave entity ubiquitous in $\mathbb{R}_{1,3}$. Therefore, all the above-discussed problems are specific to the quantum picture.

2.6.2. *Motions of a Particle Along a Straight Line*

Before proceeding to problems on one-dimensional motions of a particle, we express the four-velocity $v^\mu = dz^\mu/ds$ and the four-acceleration $a^\mu = dv^\mu/ds$ in terms of their three-dimensional vector components $\mathbf{v}$ and $\mathbf{a}$ in a particular Lorentz frame. With reference to Eq. (16), we find $dt/ds = \gamma$ and $d\gamma/dt = \gamma^3\,(\mathbf{a}\cdot\mathbf{v})$, so that

$$v^\mu = (\gamma, \gamma\,\mathbf{v}), \quad a^\mu = \left((\mathbf{a}\cdot\mathbf{v})\,\gamma^4, \, \mathbf{a}\,\gamma^2 + \mathbf{v}\,(\mathbf{a}\cdot\mathbf{v})\,\gamma^4\right). \tag{65}$$

In the instantaneously comoving frame, $\mathbf{v} = \mathbf{0}$, and

$$v^\mu = (1, \, \mathbf{0}), \quad a^\mu = (0, \, \mathbf{a}), \tag{66}$$

therefore,

$$a^2 = -\mathbf{a}^2. \tag{67}$$

The acceleration magnitude $|\mathbf{a}|$ as seen in an instantaneously comoving frame proves to be a Lorentz invariant equal to $\sqrt{-a^2}$.

Let a particle be moving along a straight line, say the x axis. For such a movement, the identity (17) takes the form: $v_0^2 - v_1^2 = 1$. Comparing it with the identity

$$\cosh^2\alpha - \sinh^2\alpha = 1, \tag{68}$$

we obtain

$$v^\mu = (\cosh\alpha, \, \sinh\alpha, \, 0, \, 0), \tag{69}$$

where α is an arbitrary function of s. We differentiate Eq. (69) with respect to s to give

$$a^\mu = \dot{\alpha}\,(\sinh\alpha, \, \cosh\alpha, \, 0, \, 0), \tag{70}$$

where the overdot denotes the derivative with respect to s. From Eq. (70) it follows that

$$a^2 = -\dot{\alpha}^2\,. \tag{71}$$

We thus conclude that any one-dimensional motion is entirely characterized by the scalar function $\alpha(s)$. Let α satisfy the differential equation

$$\ddot{\alpha} = 0\,, \tag{72}$$

whose integration gives

$$\alpha(s) = \alpha_0 + ws\,, \tag{73}$$

where α_0 and w are arbitrary integration constants. In view of Eq. (71), we have

$$a^2 = -w^2\,. \tag{74}$$

Comparison of Eqs. (74) and (67) shows that Eq. (73) describes the movement which is naturally called uniformly accelerated. To be more specific, a particle executes relativistic *uniformly accelerated* motion if its acceleration $\mathbf{a}$ in the instantaneously comoving frame remains the same at any instant. We will see below that this definition is consistent with the usual concept of uniformly accelerated motion in Newtonian mechanics when $|v| \ll 1$.

For simplicity, we set $\alpha_0 = 0$ in Eq. (73). Then substituting (73) into (69), we obtain

$$v^\mu(s) = (\cosh ws,\, \sinh ws,\, 0,\, 0)\,. \tag{75}$$

Integration of Eq. (75) over s gives the world line of a uniformly accelerated particle:

$$z^\mu(s) = z^\mu(0) + \frac{1}{w}\,(\sinh ws,\, \cosh ws,\, 0,\, 0)\,. \tag{76}$$

From this expression it is easy to read how the coordinates t and x depend on s:

$$t - t_0 = \frac{1}{w}\sinh ws\,, \quad x - x_0 = \frac{1}{w}\cosh ws\,. \tag{77}$$

It follows the trajectory

$$x - x_0 = \frac{1}{w}\,\sqrt{1 + w^2(t - t_0)^2}\,. \tag{78}$$

The curve defined by Eq. (76) is a *hyperbola* which asymptotically approaches the light cone ray $t + x = 0$ as $t \to -\infty$, and another ray $t - x = 0$ as $t \to \infty$, Fig. 1, hence the name *hyperbolic motion*. This term is used in the literature as a synonym for the relativistic uniformly accelerated motion.

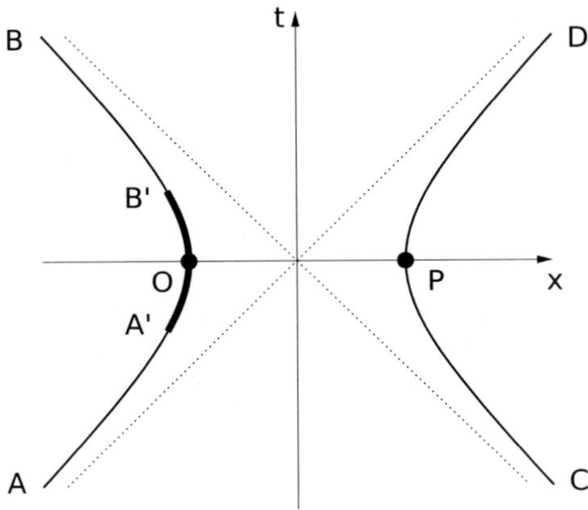

Figure 1. World lines of a hyperbolic motion.

Differentiating Eq. (78), we find the velocity of the particle in the x-axis direction,

$$v = \frac{w\,(t - t_0)}{\sqrt{1 + w^2(t - t_0)^2}}\,. \tag{79}$$

The velocity v is small, $|v| \ll 1$, during the period $|t - t_0| \ll w^{-1}$. In Fig. 1 this is a fragment of the hyperbola, $A'B'$, close in appearance to a segment of a vertical straight line. During this period, Eqs. (78) and (79) give the well-known non-relativistic formulas for uniformly accelerated motions: $x = X_0 + \frac{1}{2}\,w\,(t - t_0)^2$ and $v = w\,(t - t_0)$.

If the speed of light is $c = 1$, then the length is expressed in units of time, say in seconds; the velocity is dimensionless; and the dimension of the acceleration is inversely related to time. The reciprocal of the free fall acceleration, g^{-1}, is equal to one year with an accuracy of 3%. This fact is convenient to use in problems concerning long space travels because dimensionless quantities like wt occur frequently in calculations. Consider two typical problems. Let the rocket engines furnish a constant acceleration $1g$ relative to the instantaneously comoving frame. How far will the rocket move away from the Earth during 40 years measured by a clock on the Earth? How much time measured by a clock attached to the rocket will it take for the astronaut to arrive at the 'edge of the Universe', that is, to cover a distance of $1.4 \cdot 10^{10}$ light years, and return back to the Earth?

Note that the astronaut feels the usual gravity of the Earth, $1g$, and we take the opportunity to put $w = 1$ in Eqs. (78) and (76), and measure t, s, and x in years.

To answer the first question, we use Eq. (78) and take into account the initial condition $x|_{t=0} = 0$,

$$x = w^{-1}\sqrt{1 + w^2t^2} - w^{-1} = \sqrt{1 + 40^2} - 1 \approx 39\,. \tag{80}$$

The rocket will move away from the Earth at a distance of 39 light years.

To answer the second question, we observe that the pertinent world line consists of four almost smoothly joined fragments of equal length (acceleration + deceleration on the way to the 'edge of the Universe', that is, the fragments PD and AO of the hyperbola in Fig. 1, and the same acceleration + deceleration on the way back, that is, the fragments OB and CP of the hyperbola), each being described by Eq. (76). It is convenient to use the relation $\sinh^{-1} x = \ln\left(x + \sqrt{x^2 + 1}\right)$ and take into account that the required time s corresponds to movement along a path of twice the length of $x = 1.4 \cdot 10^{10}$ light years,

$$s = \ln\left(2x + \sqrt{(2x)^2 + 1}\right) \approx \ln\left(5.6 \cdot 10^{10}\right) \approx \ln 5.6 + 10 \ln 10 \approx 1.7 + 23 \approx 25\,. \tag{81}$$

The astronaut will spend 25 years of his life for the travel to the 'edge of the Universe' and back to the Earth, whereas a clock on the Earth will read $2.8 \cdot 10^{10}$ years.

Returning to the first problem, we can similarly find that a clock of the astronaut will read 4.4 years in his travel at a distance of 39 light years away from the Earth.

Taken together these problems offer a good illustration of the 'twin paradox'. Figure 1 may be useful in the treatment of other paradoxes, such as 'Bell's spaceship paradox'.

2.6.3. *A Charged Particle in Homogeneous Constant Electromagnetic Fields*

The behavior of a charged particle in an electromagnetic field is governed by the equation of motion (26) in which f^μ is the Lorentz four-force (33),

$$m\,\frac{dv^\mu}{ds} = qv_\nu F^{\mu\nu}\,. \tag{82}$$

If the particle is free, $F^{\mu\nu} = 0$, Eq. (82) predicts a straight world line, $z^\mu(s) = z^\mu(0) + V^\mu s$. This fact can be taken as another definition of a Galilean particle [11]. A straight line is a well-defined geometric concept in $\mathbb{R}_{1,3}$. By contrast, expressing the concept of free motions in terms of variables peculiar to Newtonian mechanics

[11] Non-Galilean free particles can behave in a different fashion: oscillate about a straight world line, and execute self-accelerated or self-decelerated motions [12].

is not quite simple. The motion of a free particle is rectilinear and *uniform* provided that the clock reads the *absolute* Newtonian time t. If another time scale, defined as a nonlinear function of the absolute time $\tau = \tau(t)$, is used, then the motion parameterized by τ is no longer uniform.

Let the field $F^{\mu\nu}$ be constant and homogeneous. Because Eq. (82) is a differential first-order equation with constant coefficients, the solution to the Cauchy problem for this equation *linear* in v^μ is almost obvious [25]:

$$v^\mu(s) = \Lambda^{\mu\nu}\, v_\nu(0) \equiv \exp\left(\frac{q}{m}\, F^{\mu\nu} s\right) v_\nu(0)\,. \tag{83}$$

For the reader aware of rudiments of the Lie algebra of the Lorentz group, this solution will resemble the action of an element of the Lorentz group $\Lambda^{\mu\nu}$ on the initial four-velocity $v^\mu(0)$, with $F^{\mu\nu}$ being the 'generator' of the Lorentz transformation $\Lambda^{\mu\nu}$. If this generator is expressed in terms of $\mathbf{E}$ and $\mathbf{B}$, then $\mathbf{E}$ generates Lorentz boosts, and $\mathbf{B}$ generates three-dimensional rotations.

Any movement of a particle draws a curve on the upper part of the hyperboloid $v^2 = 1$. For a homogeneous constant field $F_{\mu\nu}$, the form of the curve can be seen directly from Eq. (83). It is the intersection of the hyperboloid and a two-dimensional plane associated with the 2-form $F = F_{\mu\nu} dx^\mu \wedge dx^\nu$. For example, if $F^{\mu\nu}$ is represented by a pure electric field F^{01} aligned with the x axis in a particular Lorentz frame, then the intersection is a hyperbola, Eq. (75), that is, we are dealing with the hyperbolic motion, Eq. (76). If $F^{\mu\nu}$ is represented by a magnetic field F^{32} directed along the z axis, then the intersection is an ellipse, and hence, the world line is a timelike helix.

The reader can find out other intersections of the hyperboloid and the two-dimensional planes associated with homogeneous constant fields, which makes it possible to reconstruct the corresponding world lines.

When this idyllic four-dimensional affair is compared with the techniques designed to integrate the Poincaré–Planck equation of motion for a relativistic particle [12] it is sufficient to note that Eq. (15) is an essentially *nonlinear* differential equation no matter what the right side may be.

3. Electrodynamics

The acquaintance of students with Maxwell's equations occurs in either of two ways. The Jackson textbook [19] presents these equations using the inductive

[12] Unfortunately, this archaic machinery still dominates the course of theoretical physics, see, *e. g.*, [13].

method, while in the Landau and Lifshitz course of theoretical physics [13], they are obtained deductively from the principle of least action. Both ways are logically perfect. However, none of them gives an insight into what is encoded in these equations.

In the 19th century, people tried to comprehend electrodynamics from the mechanics of an aether, guided by an illusory analogy with hydrodynamics. Today we know that Maxwell's equations are largely to be a reflection of the properties of space and time. However, does their content boil down to stating that the room for events is $\mathbb{R}_{1,3}$, or are they talking about something else?

3.1. Geometric Content of Maxwell's Equations

Our concern here is to understand to what extent the form of the dynamical law for the electromagnetic field is ordered by the geometrical features of our world, specifically by the fact that space is three-dimensional and Euclidean.

Any smooth vector function $\mathbf{V}(\mathbf{x})$ can be reconstructed from 9 components of its gradients $\nabla_i V_j$. In fact, to do this requires much less information. The Helmholtz theorem [26] states that if $\mathbf{V}(\mathbf{x})$ disappears at infinity, then it can be uniquely restored from its curl $\mathbf{C} = \nabla\times\mathbf{V}$ and divergence $D = \nabla\cdot\mathbf{V}$, that is, from 4 components of linear combinations of $\nabla_i V_j$ [13]. The next point is that the electromagnetic field is a physical object whose states are described by an antisymmetric tensor $F^{\mu\nu}$. We now assume that the law of evolution of the electromagnetic field is given by a set of partial differential equations for $F^{\mu\nu}$. It follows from the Helmholtz theorem that these equations do not necessarily contain all 24 components of the derivatives $\partial_\lambda F_{\mu\nu}$, and it will suffice to involve only those linear combinations of them which are expressed in terms of curls and divergences of $\mathbf{E}$ and $\mathbf{B}$. There are 8 components of such linear combinations:

$$\partial_\mu F^{\mu\nu} = \left(\operatorname{div}\mathbf{E}, -\frac{\partial\mathbf{E}}{\partial t} + \operatorname{curl}\mathbf{B}\right) \tag{86}$$

and

$$\partial_\mu {}^*F^{\mu\nu} = \left(\operatorname{div}\mathbf{B}, -\frac{\partial\mathbf{B}}{\partial t} - \operatorname{curl}\mathbf{E}\right). \tag{87}$$

[13]The proof of the Helmholtz theorem is simple. The relationship

$$\nabla\times(\nabla\times\mathbf{V}) = \nabla(\nabla\cdot\mathbf{V}) - \nabla^2\mathbf{V}, \tag{84}$$

which is well known from elementary vector analysis, can be rewritten as the Poisson equation,

$$\nabla^2\mathbf{V} = \mathbf{S}, \tag{85}$$

with $\mathbf{S}$ being a given function $\mathbf{S} = \nabla D - \nabla\times\mathbf{C}$. The extremum principle for harmonic functions holds that the solution to Eq. (85) is unique provided that $\mathbf{V}(\mathbf{x})\to 0$ as $|\mathbf{x}|\to\infty$.

The origin of the differential operations 'curl' and 'div' becomes clear. They have nothing to do with vortices, sources and sinks of quantities intended for the description of hydrodynamics of the aether. From essentially geometric considerations the following field equations arise

$$\partial_\lambda F^{\lambda\mu} = 4\pi j^\mu\,, \tag{88}$$

$$\partial_\mu {}^*F^{\mu\nu} = 4\pi m^\mu\,, \tag{89}$$

where j^μ and m^μ are the sources of local changes in the state of the electromagnetic field. The construction of the four-vectors j^μ and m^μ is at this stage indeterminate.

3.2. Physical Content of Maxwell's Equations

It remains to clarify what j^μ and m^μ are . To do this requires three additional assumptions:

i) *Linearity* of the field equations, or the *superposition* principle;

(ii) The extended *action–reaction* principle;

(iii) *Lack of magnetic monopoles.*

Let N charged particles interacting with the electromagnetic field form a closed system. The left sides of Eqs. (88) and (89) are linear in $F^{\mu\nu}$ to fit these equations in assumption (i). The explicit Lorentz covariance can be used to conclude that j^μ and m^μ are independent of $F^{\mu\nu}$. The quantities j^μ and m^μ may depend on particle characteristics such as coupling constants q_I and world line variables $z_I^\mu(s_I)$. What are those dependencies? If we apply the operator ∂_μ to Eq. (88), we obtain $\partial_\mu\partial_\lambda F^{\lambda\mu} \equiv 0$ because $F^{\lambda\mu}$ is antisymmetric in λ and μ, and $\partial_\mu\partial_\lambda$ is symmetric in these indices. Therefore, to ensure the consistency of Eq. (88), the relation

$$\partial_\mu j^\mu = 0 \tag{90}$$

must hold identically, namely for any value of q_I and any function $z_I^\mu(s_I)$. Integrating Eq. (90) over a region of $\mathbb{R}_{1,3}$ bounded by a timelike tube of large radius and two spacelike hypersurfaces Σ_1 and Σ_2, and applying the Gauß–Ostrogradskiĭ theorem, we verify that the quantity

$$Q = \int_\Sigma d\sigma_\mu\, j^\mu \tag{91}$$

is independent of the shape and position of the integration surface Σ. In particular, Q remains unchanged under the shift of Σ along the time axis. The quantity Q is called the total *charge-source*. The constancy of the charge-source, $Q = \text{const}$, can reasonably be related to the constancy of the charge-coupling q implied by Eq. (34). With reference to the fact that the hypersurface Σ is intersected by N world lines of charged particles, assumption (ii) can be made concrete:

$$Q = \sum_{I=1}^{N} q_I \,. \tag{92}$$

Imagine for a while that only a single particle with the coupling q is in the Universe, then

$$Q = q\,. \tag{93}$$

This equation means the identity of the charge-source and the charge-coupling. Indeed, it is just the extended action–reaction principle: the charge-coupling measures the variation of the particle state for a given electromagnetic field state while the charge-source measures the variation of the electromagnetic field state for a given particle state. Therefore, both quantities could be reasonably lumped together as the *electric charge* or briefly the *charge*.

How can Eq. (92) be implemented technically? With this aim in mind let us write j^μ in the form

$$j^\mu(x) = \sum_{I=1}^{N} q_I \int_{-\infty}^{\infty} ds_I\, v_I^\mu(s_I)\, \delta^{(4)}\big[x - z_I(s_I)\big]\,. \tag{94}$$

It is readily shown that this j^μ, called the electric *current density*, obeys Eq. (90). Hence, the integration surface Σ in Eq. (91) is arbitrary. We define it such that all the world lines are perpendicular to Σ at the intersection points. For a small vicinity of the intersection point, we have $ds_I\, d\sigma_\mu v_I^\mu = d^4x$, where x^μ is a Cartesian coordinate in the Lorentz frame with the time axis directed along v_I^μ. Inserting (94) into (91) gives (92).

We next turn to Eq. (89). We reiterate *mutatis mutandis* the above arguments to conclude that m^μ is independent of $F^{\mu\nu}$, but may depend on particle characteristics. The comparison between Eqs. (86) and (87) shows that the roles of the electric and magnetic fields are now interchanged. Therefore, only particles possessing magnetic couplings $q_I^\star$ contribute to m^μ. In line with the extended action–reaction principle, the total magnetic charge-source $Q^\star$, defined as

$$Q^\star = \int d\sigma_\mu\, m^\mu\,, \tag{95}$$

equals the sum of magnetic charge-couplings,

$$Q^\star = \sum_{I=1}^{N} q_I^\star\,. \tag{96}$$

With assumption (iii), we culminate in

$$m^{\mu} = 0\,. \tag{97}$$

These observations lead us directly and unambiguously down to Maxwell's equations (1) and (2). If we take into consideration Eqs. (86) and (87), and write the current density components $j^{\mu} = (\varrho, \mathbf{j})$, we arrive at Maxwell's equations in three-dimensional vector form, Eqs. (3) and (4).

Partial differential equations of the hyperbolic type are best suited for the expression of the *local interaction* paradigm. Maxwell's equations fall in this category. An alternative concept, the so-called *action-at-a-distance*, will be outlined in subsection 3.7..

To summarize, a major part of the information encoded in Maxwell's equations is taken from global topological properties of $\mathbb{R}_{1,3}$. The residual information, seemingly divorced from geometry, represents the physical contents of these equations, which translates into three assumptions (i)–(iii).

3.3. Linearity vs Nonlinearity

Turning to the above line of reasoning one may ask: Are assumptions (i)–(iii) essential for displaying the physical content of Maxwell's equations? Is it actually impossible to replace some of them with another assumption, say, that reflecting the symmetry properties of Maxwell's equations? This issue is rather extended. We address it here, and continue our analysis it in the following subsections.

One may wonder whether the equations of motion for the electromagnetic field are linear in the strict sense or if this is merely a good approximation for the phenomena we have so far been able to observe. It is then possible to extend the previous framework for description of the electromagnetic field by discarding linearity but retaining locality and the absence of magnetic monopoles, and proceed from the Lagrangians that depend on the field strength expressed in terms of smooth vector potentials, $F_{\mu\nu} = \partial_{\mu}A_{\nu} - \partial_{\nu}A_{\mu}$. Although this consideration is limited to Lagrangians which are nonlinear functions of the electromagnetic field invariants $\mathcal{S} = \frac{1}{2}F_{\mu\nu}F^{\mu\nu}$ and $\mathcal{P} = \frac{1}{2}F_{\mu\nu}{}^{*}F^{\mu\nu}$, such extensions of the Maxwell–Lorentz electrodynamics are many and varied. We thus have to restrict our attention to the two best-studied examples.

Imposing additional constraints (not too convincing from the present standpoint), Max Born and Leopold Infeld proposed a version of nonlinear electrodynamics [27] which held much promise. This theory turned out to be mathematically complicated, and soon disappeared from the agenda. Nevertheless, half a century later,

Efim Samoĭlovich Fradkin and Arkadiĭ Aleksandrovich Tseĭtlin revived interest in the Born–Infeld electrodynamics, showing that it arises as an effective theory in the low-energy limit of string theory [28].

Maxwell's equations are invariant not only under the 10-parameter Poincaré group of Lorentz transformations and translations, but under the larger, 15-parameter conformal group of spacetime transformations $\mathrm{C}(1,3)$. This fact was discovered by Harry Bateman [29] and Ebenezer Cunningham [30] in 1909. Since then the conformal symmetry has become one of the most intriguing research topics in field theory.

Why is this symmetry so attractive? Mathematically, the conformal group $\mathrm{C}(1,3)$ is the lowest dimensional group containing the Poincaré group. Of special note is that $\mathrm{C}(1,3)$ is semisimple, even though the Poincaré group is the semidirect product of the Lorentz and translation groups. Conformal invariance is of significant physical interest because it is the maximal spacetime symmetry of Maxwell's equations [31]. Furthermore, Maxwell's equations enjoy the property of conformal invariance only in spacetime of dimension $D = 4$. In contrast, the Born–Infeld theory it is not invariant under scale transformations, and hence devoid of conformal symmetry.

Another important property of Maxwell's equations is that they are dual of electric and magnetic fields. The idea of duality goes back to Oliver Heaviside [32], who observed that $\mathbf{E}$ and $\mathbf{B}$ enter Maxwell's equations in a symmetric way. Joseph Larmor noticed [33] that these equations with $\rho = 0, \mathbf{j} = 0$ are invariant under the discrete transformations $\mathbf{E} \to \mathbf{B}$ and $\mathbf{B} \to -\mathbf{E}$. Yuriĭ Germanovich Raĭnich showed [34] that Maxwell's equations without sources are invariant under the electric-magnetic duality rotation

$$\begin{aligned} \mathbf{E}' &= \mathbf{E}\cos\theta + \mathbf{B}\sin\theta\,, \\ \mathbf{B}' &= -\mathbf{E}\sin\theta + \mathbf{B}\cos\theta\,. \end{aligned} \tag{98}$$

There is a generalization of rotations (98) for nonlinear variants of electrodynamics [35]. Ivo Bialynicki-Birula established that the Born–Infeld theory has this dual property [35].

Hermann Weyl believed that Maxwell's theory stands out against nonlinear extensions of electrodynamics in that it is the only theory which is endowed with conformal symmetry [36]. However, this hypothesis did not prove true. One may further assume that Maxwell's electrodynamics is the only theory invariant under transformations of the $\mathrm{C}(1,3)$ group and duality rotations. Were this indeed the case, the linearity of Maxwell's equations would be tantamount to the availability of these symmetries. This assumption was also disproved. There is a (unique) nonlinear extension of Maxwell's electrodynamics that has these maximum allowable symmetries [37], [38]. This theory is referred to as *ModMax*.

3.4. Coexistence of Electric and Magnetic Charges

If assumption (iii) is abandoned, the dynamical law for the electromagnetic field takes the form of Eqs. (88) and (89). These equations are invariant under the duality rotations (98) despite the fact that both sources j^μ and m^μ are nonvanishing. With the designations

$$\mathcal{F}^{\mu\nu} = F^{\mu\nu} - i\,{}^*F^{\mu\nu}\,, \quad \mathcal{J}^\mu = j^\mu - im^\mu\,, \tag{99}$$

Eqs. (88) and (89) are converted to a single complex-valued equation

$$\partial_\lambda \mathcal{F}^{\lambda\mu} = 4\pi \mathcal{J}^\mu\,, \tag{100}$$

invariant under the electric-magnetic duality transformations

$$\mathcal{F}'_{\mu\nu} = \mathcal{F}_{\mu\nu}\, e^{i\theta}\,, \quad \mathcal{J}'_\mu = \mathcal{J}_\mu\, e^{i\theta}\,. \tag{101}$$

Although Eq. (100) acquires additional symmetry, shown in Eq. (101), the theory as a whole is more involved than the Maxwell–Lorentz electrodynamics because the concept of smooth vector potentials A_μ falls victim to the electric-magnetic duality.

Within about 100 years, the idea of magnetic charges was banished from the physical scene. The troublemaker was Paul Adrien Maurice Dirac [39], [40] who established the connection between the existence of a magnetic monopole and discreteness of electric charges. Dirac retained the concept of vector potentials but was forced to postulate a singular string coming out from the point where the monopole is located and going to infinity. The idea is that the divergence of $\mathbf{B}$ vanishes almost everywhere even though the relation $\mathbf{B} = \operatorname{curl} \mathbf{A}$ holds almost everywhere, and yet the total flux of $\mathbf{B}$, concentrated at the singular line, is $4\pi q^\star$. The string may be conceived as a thin solenoid that carries the magnetic flux $4\pi q^\star$ to infinity.

Another possibility to save A_μ is to use the so-called Wu–Yang vector potential [41] which maps $\mathbb{R}_{1,3}$ onto a four-dimensional manifold with more complicated topology. In any case, A_μ is an unattractive construction whose aesthetic damage is to be compensated by Dirac's explanation of where the *elementary* electric charge comes from. No other convincing explanation has yet been proposed [42].

The simplest way to show the discreteness of electric charges is as follows. Poincaré established that the conservation law for angular momentum $\mathbf{L}$ is violated in the binary system composed of a monopole and a charged particle [43] . The violation is due to the fact that only mechanical part of the total angular momentum $\mathbf{J}$ is taken into cosideration. If one takes proper account of the contribution of electromagnetic field, $\mathbf{l} = -qq^\star \mathbf{n}$, where q is the electric charge of the particle, $q^\star$ the magnetic charge of the monopole, and $\mathbf{n}$ a unit vector directed from the magnetic monopole to the charged particle, then the conservation law for the total angular momentum $\mathbf{J} = \mathbf{L} - qq^\star\mathbf{n}$ is obeyed [44]. In quantum mechanics, the projection of angular momentum $\mathbf{l}$ onto any axis takes values which are multiples of $\frac{1}{2}$ (in terms of $\hbar$). Therefore, the expression $\mathbf{l} = -q\,q^\star\mathbf{n}$ enables us to explain the quantization of the electric charge: If there exists at least one monopole in the Universe, then the following relation holds:

$$q\,q^\star = \frac{1}{2}\,n\,, \quad n = 1, 2, \ldots\,, \tag{102}$$

which evidences that q is a multiple of $(2q^\star)^{-1}$.

3.5. Self-interaction

Turning to assumption (ii), two comments are in order. Firstly, this assumption renders distinct some value of spacetime dimension D. Indeed, the state of the electromagnetic field is characterized by $\frac{1}{2}\,(D-1)\,D$ components of the field strength $F^{\mu\nu}$, while that of a charged point particle is characterized by $2\,(D-1)$ coordinates of its phase space. The action–reaction principle requires that these quantities be equal. This gives $D = 4$ [14].

It transpires the sense of the generalized action–reaction principle. There are two structure elements of classical reality, *particles* and *fields*. If they are represented by the charged particle and the electromagnetic field, it is imperative that spacetime be $\mathbb{R}_{1,3}$; otherwise those structure elements will not go well together.

Secondly, assumption (ii) is selective for the nature of the dynamics. The action–reaction principle is applicable to electrodynamics and Yang–Mills theory, but has no relation to the theory of gravity [10]. To verify this, we note that gravity is exerted uniformly on any particle regardless of its mass, namely the equation of motion for a particle in gravitational fields, the geodesic equation (54), is free from the particle mass m. In contrast, particles with different masses affect the gravitational field in different ways. Consider the equation of motion for the gravitational field

$$R^{\mu\nu} - \frac{1}{2}\,R\,g^{\mu\nu} = 8\pi G_{\rm N} T^{\mu\nu} \tag{103}$$

with the stress-energy tensor of a single particle endowed with mass $m_{\rm g}$,

$$T^{\mu\nu} = m_{\rm g}\int_{-\infty}^{\infty} ds\, v^\mu(s)\, v^\nu(s)\, \delta^{(4)}[x - z(s)]\,. \tag{104}$$

[14]The solution $D = 1$ holds no physical value. There is no pictorial rendition for all that may occur in one-dimensional worlds.

Clearly, the larger is $m_{\rm g}$, the stronger is the effect on the gravitational field characterized by $R^{\mu\nu} - \frac{1}{2}\,R\,g^{\mu\nu}$. We thus see that the action–reaction principle is incompatible with Einstein's principle of equivalence, Eq. (46).

The action–reaction principle in Maxwell–Lorentz electrodynamics is most pronounced with the problem of self-interaction [45]. In a naive interpretation, this problem is to find the interaction of a point particle with its own field. However, the field generated by this particle is singular on its world line, and hence all measured physical quantities, such as energy-momentum and angular momentum, are given by mathematically meaningless divergent expressions. Using regularization and renormalization methods it is possible to find finite, uniquely defined expressions for these quantities, as well as the equation of motion for a particle as regards self-interaction, the Abraham–Lorentz–Dirac equation [46], [47], [48],

$$ma_\lambda - \frac{2}{3}q^2\left(\dot a_\lambda + a^2 v_\lambda\right) = qv^\mu F^{\rm ext}_{\lambda\mu}(z)\,. \tag{105}$$

Here, m is the renormalized mass, a finite sum of the 'bare particle mass' m_0 and the electromagnetic field self-energy,

$$m = \lim_{\epsilon\to 0}\left[m_0(\epsilon) + \frac{q^2}{2\epsilon}\right], \tag{106}$$

and $F^{\rm ext}_{\lambda\mu}(z)$ is an external electromagnetic field.

A closer look at self-interaction shows that it can be understood as a rearrangement of the initial degrees of freedom of a bare particle and an amorphous electromagnetic field, appearing in the action of the Maxwell–Lorentz theory, to convert these entities into a *dressed* particle and *radiation* [49]–[52]. Equation (105) is thus merely the equation of relativistic dynamics (23) for a dressed particle. Indeed, on substitution of the expression for the four-momentum of a dressed particle, derived by Claudio Teitelboim [53],

$$p^\mu = mv^\mu - \frac{2}{3}\,q^2 a^\mu\,, \tag{107}$$

into Eq. (23), and taking into account the identities $v\cdot a = 0$ and $v\cdot\dot a = -a^2$ which follow from differentiation of Eq. (17), we come to Eq. (105).

On the other hand, Eq. (105) is the energy-momentum balance on the world line of the dressed particle: the four-momentum $d\wp^\lambda = qF^{\lambda\mu}_{\rm ext}\,v_\mu ds$, extracted from the external field $F^{\lambda\mu}_{\rm ext}$ over an infinitely short period of time ds, is spent on the increment of the dressed particle four-momentum dp^λ, and the four-momentum carried away by radiation $d{\cal P}^\lambda$,

$$dp^\lambda + d{\cal P}^\lambda = d\wp^\lambda\,. \tag{108}$$

To see this, we substitute the increment of the four-momentum p^μ of the dressed particle defined by Eq. (107), the external field four-momentum $d\wp^\lambda$, that is, the Lorentz force $qF^{\lambda\mu}_{\rm ext}\,v_\mu ds$ acting for an infinitely short period of time ds, and the differential of the four-momentum $\mathcal{P}^\lambda$ equal to the Larmor radiation intensity [54], [55],

$$\dot{\mathcal{P}}^\mu = -\frac{2}{3}\,q^2a^2v^\mu\,, \tag{109}$$

multiplied by ds, in Eq. (108), to regain Eq. (105).

If the external field is absent, then the variation of the four-momentum of the dressed particle is the response to the four-momentum carried away by radiation, $dp^\mu = -d\mathcal{P}^\lambda$. Perhaps this is the most prominent manifestation of the action–reaction principle.

The salient feature of the Abraham-Lorentz-Dirac equation (105) is the absence of invariance under time reversal $s \to -s$. This is due to the fact that Eq. (105) includes the four-acceleration, which is unchanged under this transformation, $a^\mu \to a^\mu$, together with its derivative $\dot{a}^\mu$, transforming as $\dot{a}^\mu \to -\dot{a}^\mu$. After rearranging the degrees of freedom, the Maxwell-Lorentz theory loses its time-reversal symmetry. Radiation is an unidirectional process, and the equation of motion for a dressed particle (105) reflects the irreversibility of this process.

A similar situation occurs in the SU($\mathcal{N}$) Yang–Mills–Wong theory [56], which describes a system of K point particles ($\mathcal{N} \ge K+1$), carrying non-Abelian charges and interacting with the Yang–Mills field. However, there is some difference. In the non-Abelian phase, a particle accelerated by an external four-force f^μ does not emit, but *absorbs* the Yang–Mills field energy, $\dot{\mathcal{P}}\cdot v < 0$, and so the equation of motion for a dressed particle is

$$m\left[a^\mu + \ell\left(\dot{a}^\mu + v^\mu a^2\right)\right] = f^\mu\,, \tag{110}$$

where ℓ is a characteristic length,

$$\ell = \frac{8}{3mg^2}\left(1 - \frac{1}{\mathcal{N}}\right), \tag{111}$$

and g the Yang–Mills coupling constant. The absorption intensity and the four-momentum of the dressed particle are given, respectively, by

$$\dot{\mathcal{P}}^\mu = m\ell\, a^2\, v^\mu\,, \tag{112}$$

$$p^\mu = m\left(v^\mu + \ell a^\mu\right). \tag{113}$$

The local energy-momentum balance, defined by Eq. (108), now reads: the four-momentum $d\wp^\mu = f^\mu ds$ extracted from an external field over an infinitely small interval of time ds is distributed between the variation of the dressed particle four-momentum dp^μ and the Yang–Mills four-momentum absorption $d\mathcal{P}^\mu$. Obviously, this dynamics is also irreversible.

The self-interaction problem in general relativity is radically different from what we have seen in gauge theories. The action–reaction principle does not apply to gravity. The energy-momentum balance on the world line of a particle, Eq. (108), is no longer valid. However, the gravitational degrees of freedom are still subjected to a rearrangement which results in gravitational field configurations with non-trivial topology, as exemplified by spacetime with black holes. This dynamics is also irreversible. Once converted into a black hole, a massive star can never regain its previous state.

3.6. Is $\mathbb{R}_{1,3}$ an *a priori* Intuition?

Immanuel Kant regarded space and time as an *a priori* intuition [58], that is, innate concepts. This point of view came under a storm of all-round criticism from philosophical and scientific positions. Currently it is classified as a doctrine of only historical interest.

However, let us look at Kant's apriorism through the eyes of a modern physicist. Note that it is the *electromagnetic* picture of our world which imparts its structure elements to our mind to form the concepts of space and time. This is most easily understood when the geometric content of Maxwell's equations is taken into account. For orientation in space, we use 'navigation equipment': a visual aggregate in which images of the external world are focused by the eye's lens onto the photosensitive cells of the retina; nerve fibers that transmit retinal impulses to the central nervous system; our brain which processes these impulses; and a device that converts brain commands into mechanical movements. Any robot has a similar navigation system: radars scan space with electromagnetic pulses in the optical or infrared wavelength range; the impulses are sent via a communication channel to an on-board computer which processes the input data; and commands are issued to the motion drive device.

The main difference between these apparati is that the robot is equipped with a digital computer, while the brain can be likened to an analog computer whose purpose is to find solutions to the Cauchy problems for Maxwell's equations. These solutions are obtained by simulating the observed electromagnetic phenomena by a system of electric currents circulating in circuits of the neural network.

The ability to navigate in space and feel the passage of time are associated with a code in the analog computer that allows us to solve Maxwell's equations, accumulate and store solutions in memory, build their superpositions, transform solutions in accordance with the symmetries inherent in Maxwell's equations. This process occurs automatically and continually. Information about the properties of solutions ultimately leads *homo sapiens* to the development of the abstract concepts of *space* and *time*. This result is natural when one keeps in mind the geometric content of Maxwell's equations. If we assume that a code for incessant solving of Maxwell's equations and analyzing their properties is built into the brain [15], then this assumption is tantamount to stating that the $\mathbb{R}_{1,3}$ is an innate form of contemplation. This is the moral of the 'geometry lesson' taught to us by Kant [16].

Many animals, including octopuses, worms, and insects, have vision. Ciliates orient themselves in space. Living organisms also sense time due to the ability to discriminate between the biological existence and nonexistence, that is, existence as a dead body. Does this mean that all living entities, including viruses, have the gift of chronogeometry, being equipped with an algorithm for continually finding solutions to Maxwell's equations?

Of course, to realize the finiteness of the earthly term for each living individual is just a human capability. However, every living creature senses another important property of time – *unidirectionality* of processes in its macroscopic environment. It is this feeling that underlies the instinct of self-preservation. For this kind of sensation to arise, objective conditions are needed to compare the reversible and the irreversible. As we saw in the previous subsection, the mechanism leading to irreversibility in the classical picture is the process of radiation. As for the electromagnetic quantum picture, all processes in it are reversible; emission and absorption of a photon have equal probability amplitudes. Therefore, the computer under discussion must function in both quantum and classical modes, allowing the possibility of switching (even through 'breakdowns and repairs') from one mode to another. We then attempt to formulate a criterion for distinguishing between living and non-living:

> *A system identified with a living organism has the ability to evolve in both classical and quantum modes. The loss of this ability is its biological death.*

The course of the above reasoning is bizarre. We began with the fact that the concepts of space and time can arise in highly developed consciousness due to the imitation of the structure of $\mathbb{R}_{1,3}$ by electromagnetic processes in neural networks,

[15] Were nonlinear generalizations of Maxwell's equations installed, say the Born–Infeld equations, linear combinations of their solutions would be superfluous because they are no longer solutions; any Cauchy problem for nonlinear equations must be solved separately. The memory of characteristic electromagnetic configurations would be useless, and higher nervous activity would lose such qualities as dreams, planning, mirage, sense of proportion, and reveries. This can probably serve as an additional argument in favor of *linearity* of the laws of electrodynamics.

[16] It is time to appreciate it, even on the eve of the 300th anniversary of the birth of the great philosopher.

and ended up considering the simplest biological entities [17], protein molecules on the border between living and nonliving matter. Here the paradoxical aphorism of Karl Marx is confirmed: 'Human anatomy contains a key to the anatomy of the ape' [59].

3.7. Action at a Distance

In view of their linearity, Maxwell's equations have symmetric in time solutions suitable for description of field configurations generated by a charged particle moving along arbitrary timelike smooth world lines. To obtain such solutions we must use the sum of retarded and advanced boundary conditions invariant under the operation $t \to -t$. While on the subject of such a field generated by two charged particles, we find that the value of the field at every point is invariant under the interchange of the world lines of these particles.

A similar line of reasoning led Adriaan Fokker [60] to the discovery of action expressed in terms of particle variables,

$$S_{\rm F} = -\sum_I \int d\tau_I \left\{ m_I \sqrt{\dot z_I^2} + \frac{1}{2} \int d\tau_J \sum_{J(\neq I)} q_I q_J \dot z_I^\mu(\tau_I) \dot z_\mu^J(\tau_J)\, \delta\big[(z_I - z_J)^2\big] \right\}, \quad (114)$$

which is symmetric under interchange of past and future. This is the so-called relativistic *action-at-a-distance* theory which also bears the name of the *direct interparticle interaction* theory. Owing to the delta-function, the link between two points z_I and z_J on the Ith and Jth world lines is given by a null interval, and hence the particles at these points can be thought of as 'directly' interacting (which is a relativistic generalization of interactions by contact occurring at zero distance). Note that the Fokker action (114) involves retarded and advanced interactions on an equal footing. But there are no autonomous field degrees of freedom. It is as if particle I were affected by particle J directly, that is, without mediation of the electromagnetic field. Where did the field degrees of freedom go? John Archibald Wheeler and Richard Phillips Feynman assumed that radiation is completely absorbed by charged matter throughout the Universe [61], [62]. What is the precise formulation of this assumption? The retarded vector potential in Maxwell–Lorentz theory can be decomposed into two terms

$$A^\mu_{\rm ret} \equiv \frac{1}{2}\left(A^\mu_{\rm ret} + A^\mu_{\rm adv}\right) + \frac{1}{2}\left(A^\mu_{\rm ret} - A^\mu_{\rm adv}\right) = A^\mu_{(+)} + A^\mu_{(-)}\,. \quad (115)$$

[17] These entities, taken as *physical* systems, are extremely complex. Future research will determine whether a protein molecule exhibiting the properties of a living being is both a classical and a quantum object. In particular, how can the principle of indistinguishability of such systems with the same content and arrangement of particles (electrons and quarks assembled in atomic nuclei) be implemented?

The last term

$$A^{\mu}_{(-)} = \frac{1}{2}\left(A^{\mu}_{\rm ret} - A^{\mu}_{\rm adv}\right) \tag{116}$$

obeys the homogeneous wave equation,

$$\Box A^{\mu}_{(-)} = 0\,. \tag{117}$$

The solution to the Cauchy problem for the equation (117) with zero initial condition on a spacelike hyperplane Σ with normal n^{μ}: $A^{\mu}_{(-)}|_{\Sigma} = 0$ and $(n\cdot\partial)\,A^{\mu}_{(-)}|_{\Sigma} = 0$ is trivial,

$$A^{\mu}_{(-)}(x) = 0\,. \tag{118}$$

Let $A^{\mu}_{(-)}(x)$ be the vector potential due to all charges in the Universe. If $A^{\mu}_{(-)}$ vanishes at one time, then it is zero at all times. Wheeler and Feynman adopted Eq. (118) as a supplementary constraint to the Fokker action, and interpreted it as the *total absorption* condition. Therefore this approach is often referred to as the *absorber theory of radiation*.

Note, however, that Eq. (118) does not amount to a lack of radiation. The effect of radiation manifests itself in the local energy-momentum balance, Eq. (108), which is evidenced by the fact that the Abraham–Lorentz–Dirac equation (105) is derivable from the Fokker action (114) and the total absorption condition (118) [61], [62], [63].

The symmetry properties of the Fokker–Wheeler–Feynman theory are the same as those of Maxwell–Lorentz electrodynamics, except for invariance under the conformal group transformations. However, this symmetry can easily be restored [64] by replacing the Minkowski metric $\eta^{\mu\nu}$ used in the action (114) with the Boulware–Brown–Peccei conformal metric [65].

The action-at-a-distance theory is successfully employed in description of all classical phenomena of electromagnetism [18] and significant part of quantum phenomena [66]. This raises a question of fundamental importance: Does the electromagnetic field exist as an independent physical object with *autonomous* degrees of freedom, or is it to be regarded as a mere auxiliary mathematical concept? What is the experimental result that let us to decide between these two alternatives?

The presence of these alternatives is critically related to the fact that the field equations are linear and the feasibility to express the electromagnetic field tensor in terms of smooth vector potentials, that is, to the absence of magnetic monopoles.

It is likely that the only opportunity to refute the action-at-a-distance paradigm and confirm the reality of the electromagnetic field is related to the discovery of

[18] Except for the case that the Universe contains a single charged particle.

a magnetic monopole. Consider a system composed of a magnetic monopole and a charged particle. It was indicated in subsection 3.4. that if the dynamics of this system is expressed entirely in terms of particle variables, then the system is characterized by the angular momentum $\mathbf{L}$, which is not conserved. The extent of violation of this conservation law, $|qq^\star|$, does not depend on the distance between the charged particle and the monopole. However, taking into account the contribution from electromagnetic degrees of freedom, we get the conclusion that there is a conserved quantity, the total angular momentum $\mathbf{J} = \mathbf{L} - qq^\star\mathbf{n}$. This would strongly suggest that the electromagnetic field is real, and the action-at-a-distance paradigm is for optional use.

4. How to Marry $\mathbb{R}_{1,3}$ with Quantum Theory

In quantum mechanics, the status of time t differs drastically from that of the coordinate $\mathbf{x}$ labelling the spatial position of a particle. There is no *operator* to represent t in a Hilbert space of quantum mechanical states. Time is a mere parameter of evolution. For example, t plays this role in the Schrödinger equation. The historical fact that the Schrödinger equation was initially discovered in a non-relativistic context is of little significance. With minor modifications, this equation has successfully settled into relativistic quantum field theory, where it is usually called the Tomonaga–Schwinger equation [67]. However, in quantum field theory the coordinate $\mathbf{x}$ ceases to be an operator, returning to the role of a parameter, and forming, together with t, the room $\mathbb{R}_{1,3}$, where the drama of quantum fields is playing out. Is there a happy end of this story? Alas, questions remain.

The *modus vivendi* of a particle is constrained by the Heisenberg uncertainty relation $\Delta x \Delta p \ge \frac{1}{2}$ which is rigorously derivable if x and p are regarded as Hermitian operators satisfying the commutation relation $[x, p] = i$. Relativism requires that the uncertainty relation $\Delta t \Delta E \ge \frac{1}{2}$ also holds for time and energy, even though this relation does not follow from the basic principles, because t is not an operator. This 'uncertainty relation', necessary for the theory of measurements in quantum mechanics and the interpretation of unstable particles as resonance peaks of wave amplitudes, still remains a heuristic addition to the Heisenberg uncertainty relation. Matters cannot be improved by invoking the formalism of quantum field theory.

This example suggests that quantum physics damages the event space structure.

4.1. Allowable World Lines

Most of relativistic expressions involve the Lorentz factor γ which becomes infinite at $\mathbf{v}^2 = 1$. This precludes acceleration of particles to velocities greater than that of light. No particle can be accelerated past the light barrier. Hence, world lines with smoothly joined timelike and spacelike fragments must be eliminated from the classical picture.

Of even greater concern is the causality argument. If faster-than-light particles exist, then it is possible for an observer to send signals to his own past. A sequence of events of this kind is called a *causal cycle*. The admission of superluminal signals would result in a patently absurd process in which an observer could cause his own destruction which, in turn, prevents the destructive signal from being sent. This seems to forbid spacelike world lines from classical relativistic theory.

Is it possible to weaken the smoothness condition, allowing jumps in tangent directions of the world line?

The key point of this issue is the presence of both particles and antiparticles together with their creation and annihilation events in the relativistic context. Given a particle moving along a timelike world line oriented from the past to the future, its antiparticle may be thought of as an object identical to it in every respect but moving backward in time [68], that is, the antiparticle world line is oriented from the future to the past, as in Fig. 2, left plot.

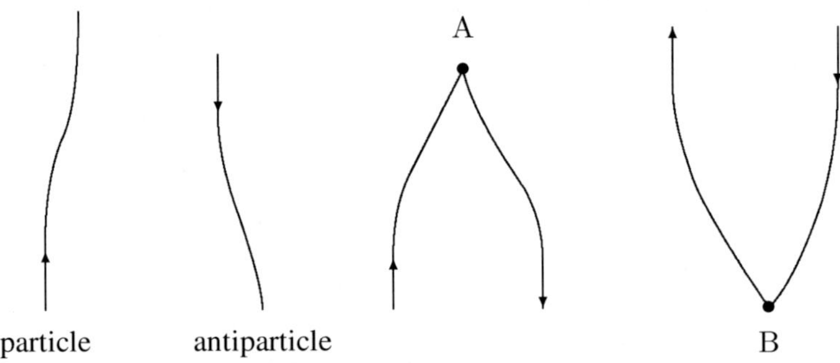

Figure 2. World lines of a particle and an antiparticle. Λ- and V-shaped paths.

Accordingly, the annihilation of a pair that occurs at a point A (A for annihilation) is depicted as a Λ-shaped world line of a *single particle* that runs initially from the remote past to the future up to the point A and then returns to the remote past. Likewise, the birth of a pair occurring at a point B (B for birth) is given by a V-shaped world line of a single particle that runs initially from the far future to the past up to the point B and then returns to the far future, as in Fig. 2, right plot.

Classical theory leaves room for both normal particles, which experience the proper order of events, and antiparticles which follow the reverse order of events. However, creations and annihilations of pairs are banned because the least action principle does not apply to V- and Λ-shaped world lines. Indeed, a spacelike hyperplane Σ may intersect a timelike V-shaped curve twice, otherwise it fails to intersect it at all. The same is true for Λ-shaped curves. Therefore, the principle of least action with endpoints separated by timelike intervals cannot be unambiguously defined for such world lines

Causal cycles can be formed from timelike Λ- and V-shaped curves. In order to exclude this possibility from the classical picture, solutions to dynamic equations should be sought within the set $\mathcal{S}_{\rm class}$ of timelike and null smooth world lines. If $\mathcal{S}_{\rm class}$ represents histories of charged particles, this results in a unique description of retarded electromagnetic fields. A charged particle moving along a world line of this kind generates a single-valued retarded field, the Liénard–Wiechert field. The restrictions imposed on $\mathcal{S}_{\rm class}$ may not be weakened, otherwise the retarded electromagnetic field is no longer single-valued.

A pictorial rendition of the classical evolution of particles is a laminar, eddy-free, flow of timelike and null smooth world lines belonging to $\mathcal{S}_{\rm class}$. It seems appropriate to call $\mathcal{S}_{\rm class}$ the *Minkowski flow*. In the quantum picture, the world line flow becomes 'turbulent'.

4.2. Geometry of Spacetime in a Quantum Context

Quantum regime of evolution allows spontaneous decays and fusions of particles, creations and annihilations of pairs. The causal cycles for virtual particles become feasible. These statements translate into the well-known criterion: the semiclassical picture is represented by *tree* diagrams of perturbation series in quantum field theory, whereas the quantum picture is displayed as *loops*.

Standard derivations of this criterion are based on a formal comparison of the powers of Planck's constant $\hbar$ that enter, as overall factors, into different terms of the perturbation series. However, the separation into trees and loops need not be related to the dilemma of whether or not the factor $\hbar^n$ presents in the expression. Mathematically, the responsibility for this separation rests with the distinction in the contents of the set of allowable world lines for a quantum-mechanical particle, $\mathcal{S}_{\rm quant}$, which involves V- and Λ-shaped curves, from the set of world lines $\mathcal{S}_{\rm class}$ free of such curves. Physically, to discriminate between loops and trees we must decide between the feasibility of creations and annihilations of pairs and veto on these processes.

A convenient framework bringing together the classical and quantum treatments is the path integral approach. Let us begin with particles. A quantum-mechanical particle can be described by the Feynman path integral [69],

$$K(\mathbf{z}_f, T|\mathbf{z}_i, 0) = \int [\mathcal{D}\mathbf{z}] \, \exp\left[i \int_0^T dt \, L(\mathbf{z}, \dot{\mathbf{z}}) \right], \tag{119}$$

where $[\mathcal{D}\mathbf{z}]$ denotes the measure of integration in the space of all paths from $\mathbf{z}_i(0)$ to $\mathbf{z}_f(T)$. Whatever the kind of $z^\mu(\tau)$ passing through the points $x^\mu = (0, \mathbf{z}_i)$ and $x^\mu = (T, \mathbf{z}_f)$ (as in Fig. 3), it makes a contribution to the Feynman path integral provided that the Lagrangian L is real, and Eq. (119) is well defined for this path.

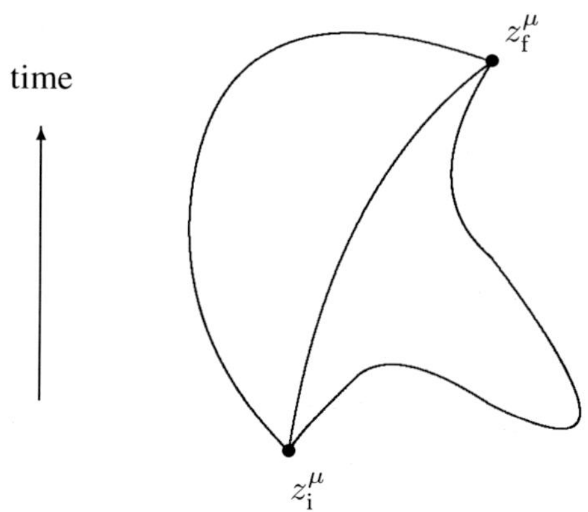

Figure 3. World lines contributing to the Feynman path integral.

To illustrate, the Poincaré–Planck Lagrangian for a free particle

$$L = -m\sqrt{\dot{z}^2} \tag{120}$$

is real and finite only for timelike paths. If $z^\mu(\tau)$ is a null curve, then $L = 0$. If $z^\mu(\tau)$ is a spacelike, then L is complex-valued. Since the imaginary part of L can take both positive and negative values, expression (119) is ill-defined. In contrast, the Lagrangian proposed in [70],

$$L = -\frac{1}{2}\left(\eta\,\dot{z}^2 + \frac{m^2}{\eta}\right), \tag{121}$$

is real and finite for timelike, null, and spacelike curves.

The semiclassical treatment implies that the extremal path contribution dominates the Feynman path integral (119). The integration over the remaining paths gives a

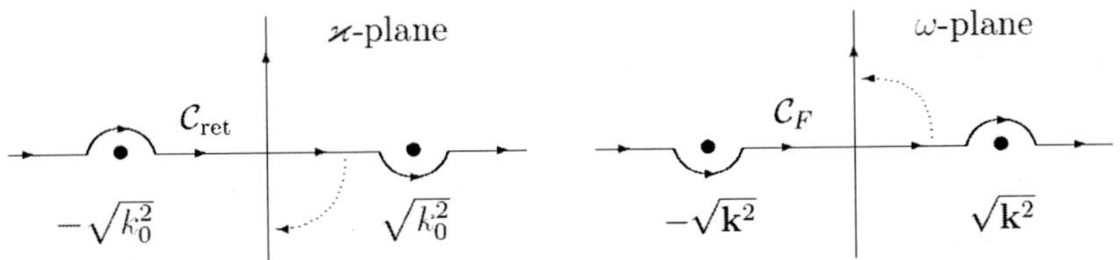

Figure 4. Integration contours $\mathcal{C}_{\rm ret}$ and $\mathcal{C}_F$ suitable for $\widetilde{D}_{\rm ret}$ and $\widetilde{D}_F$.

minor preexponential factor correction. In a strong quantum regime contributions of all paths are comparable, and in particular contributions of Λ- and V-shaped curves may be tangible.

We now turn to the field treatment. The key difference between the classical and quantum manifestations of the same field is due to the different boundary conditions imposed on their propagation laws. To be specific, consider a massless scalar field. The Fourier transform of the retarded Green's function

$$\widetilde{D}_{\rm ret}(k) = -\frac{1}{k^2 + 2ik_0\epsilon} = \frac{1}{\mathbf{k}^2 - (k_0 + i\epsilon)^2} \tag{122}$$

gives an accurate account of how this field propagates in classical theory. If the integration over the variable $\varkappa = |\mathbf{k}|$ is carried out first [19], then the poles at

$$\varkappa = \pm\sqrt{k_0^2} \pm i\epsilon \tag{123}$$

are avoided by the path depicted in the left plot of Fig. 4.

The propagation of a free massless field in quantum theory is described by the Feynman propagator

$$\widetilde{D}_F(k) = -\frac{1}{k^2 + i\epsilon} \tag{124}$$

which obeys the causal boundary condition. Then there is the prescription for avoiding the poles

$$\omega \equiv k_0 = \pm\sqrt{\mathbf{k}^2} \mp i\epsilon\,. \tag{125}$$

The integration contour of $\mathcal{C}_F$ in the complex ω-plane is depicted in the right plot of Fig. 4.

---

[19] We suppose that angular variables, if any, have already been integrated out, and that the resulting expression can be uniquely continued to the negative $\varkappa$-semiaxis.

At first glance, there is nothing in common between propagators satisfying the retarded and causal boundary conditions. This is not the case. To build a bridge between the classical and quantum descriptions of Green's functions, it is necessary to subject them to the so-called Euclideanization [20].

Assuming that the integrand decreases sufficiently fast as $\varkappa \to \infty$, it is possible to rotate the path of integration $C_{\rm ret}$ in a clockwise direction by $\frac{\pi}{2}$ in the complex $\varkappa$-plane, without crossing the poles, as in the left plot of Fig. 4. This operation, the analytical continuation to imaginary values of the complex $\varkappa$-plane, is similar to the Wick rotation. Introducing a new variable $\vec{\mathbb{K}} = i\mathbf{k}$ makes the length squared of k^μ positive definite:

$$k_E^2 = k_0^2 + \vec{\mathbb{K}}^2\,. \tag{126}$$

The analytical continuation of the space variables to the imaginary axes $\vec{\mathbb{X}} = i\mathbf{x}$, performed together with the Wick rotation in the complex $\varkappa$-plane, introduces the Euclidean metric

$$dx_E^2 = dx_0^2 + d\vec{\mathbb{X}}^2\,. \tag{127}$$

On the other hand, if we carry out the Wick rotation of the path of integration C_F in a counterclockwise direction by $\frac{\pi}{2}$ in the complex ω-plane without crossing the poles, as in the right plot of Fig. 4, which is equivalent to introducing $k_4 = ik_0$, the length squared of k^μ will become negative definite:

$$k_E^2 = -\left(k_4^2 + \mathbf{k}^2\right). \tag{128}$$

The analytical continuation of the time variable to the imaginary axis $x_4 = -ix_0$, performed together with the Wick rotation, introduces the Euclidean metric

$$dx_E^2 = -\left(dx_4^2 + d\mathbf{x}^2\right). \tag{129}$$

We thus see that we must change the overall sign of the spacetime signature in the classical description of field propagation for it to be treated as the quantum description of field propagation. Indeed, two Lorentzian metrics with opposite signatures can always be analytically continued to two Euclidean line elements of opposite sign, such as those shown in Eqs. (127) and (129).

Taken alone, the overall sign of the Lorentzian metric is of no particular importance; its choice is a matter of convention. In dynamical problems, people adopt the mainly negative signature $(+\,-\,-\,-)$, while in studying the properties of hypersurfaces, the mainly positive signature $(-\,+\,+\,+)$ is preferable. However, if this

[20] Note that this is more than a neat mathematical trick. If the Hamiltonian is bounded from below, then the analytical continuation to imaginary time can be carried out with the help of the Wick rotation [71], resulting in the Euclidean functional integral whose measure is well-defined. This tool is common practice [72], as opposed to the ill-defined Feynman path integral in its original formulation (119).

overall sign is changed as one passes from some region of spacetime to a contiguous region, then this change of sign is evidence of switching between the classical and quantum regimes of field propagation.

Such is the case for the contiguous regions inside and outside the *event horizon* of a Schwarzschild black hole. The Schwarzschild metric [13] describing an isolated spherically symmetric stationary black hole reads

$$ds^2 = \left(1 - \frac{r_S}{r}\right) dt^2 - \left(1 - \frac{r_S}{r}\right)^{-1} dr^2 - r^2 d\Omega\,. \qquad (130)$$

Here, $r_S = 2GM$ is the Schwarzschild radius which represents the event horizon of this black hole. In the Schwarzschild exterior $r > r_S$, the Killing vector field $X = \partial_t$ is interpreted as the asymptotic time translation. In the Schwarzschild interior $r < r_S$, r is a time coordinate, and the integral lines of the vector field $X = \partial_r$ are incomplete timelike geodesics which terminate at $r = 0$. Once the Euclideanization has been performed, the regions inside and outside the boundary $r = r_S$ take the Euclidean metrics of the type of (127) and (129), respectively. Following the reverse course of reasoning, we are forced to conclude that the event horizon of a Schwarzschild black hole shows a clear demarcation between spacetime regions characterized by opposite signatures. This geometric layout, if it exists, provides an explicit scheme for interfacing the classical and the quantum [73].

4.3. Space, Time, Matter, Randomness

Unlike classical realm, of which the most perfect geometric model is $\mathbb{R}_{1,3}$, the brainchild of Poincaré and Minkowski, quantum realm appears to be ‘geometrically omnivorous’. Calculations of quantum phenomena is convenient to perform after the Euclideanization, the transition from $\mathbb{R}_{1,3}$ to $\mathbb{R}_4$. To analyze the foundations of quantum field theory, it is necessary to complexify the event space, that is, to transform $\mathbb{R}_{1,3}$ into $\mathbb{C}_4$ [74], [75]. However, before judging the merits of geometric models of reality, it would be desirable to have a *general definition* of the concepts of space and time. How to formulate it?

We take, as the starting point, the following indisputable statement: physical reality is structured. We consider material objects, such as point particles, to be elements of this structure. Any object can in principle be thought of as existing or as nonexistent. However, moving from logic to reality, we find that the symmetry between existence and non-existence is broken. If a particle does not exist, then this fact is clear and unambiguous. On the other hand, if a particle exists, then this can be understood in two ways: either as the presence of the particle in a given place and at a given instant or as its implicit existence. The intervention of the clarifications ‘here’ and ‘now’ leads to the conclusion that the existence of a particular object and its non-existence are concepts of different weights. Hence, it seems reasonable to infer

> *Space and time are concepts that express the violation of symmetry between the existence and non-existence of structural elements of physical reality.*

This inference limits the ontological content of the defined concepts. As for geometric models, it is difficult to separate objective facts from peculiarities of our mental processes and the circumstances of the history of knowledge. For centuries people believed in ancient geocentrism; it was followed by the era of Newtonian absolute space and absolute time; and, finally, our mind was captivated by the Minkowski world $\mathbb{R}_{1,3}$, which opened up new depths of classical realm.

The proposed definition of space and time implies the clearly fixed ontological status of any material object; a particle may either be granted existence or not – *tertium non datur.* But the structure of reality does not have to exhibit existential certainty. The existence of structure elements can be *random.* This is exactly what quantum reality is. Its elements are still called particles, but from the point of view of a classical observer, the behavior of such particles obeys *probabilistic* laws. Particles are capable of *spontaneously* converting into other particles. For example, a d quark is prone to β-decay:

$$d \to u + e + \bar{\nu}_e \, . \tag{131}$$

The state of a d quark is a superposition of the states of an undecayed d quark and its decay products, that is, at any instant, there is some probability W that this d quark still exists, but with probability $1 - W$ it no longer exists.

The lack of existential certainty is also typical for stable quantum objects, such as electrons. Unlike a classical particle, which retains its individuality throughout its entire history, the electron does not care about 'saving face' at all. It is identical to any electron in any corner of the Universe, even to the electron that has not yet been born in the reaction (131), and is fundamentally indistinguishable from it.

In order to physically implement a probabilistic law governing the random behavior of a certain system, it is necessary to have a *statistical ensemble* of such systems [76], [77]. And since our concern is with systems whose identical and indistinguishable elements occupy the entire spacetime, we must find out what is the rationale for assuming the existence of an ensemble of such arenas.

In the early 80s of the last century, Michael Freedman [78] and Simon Donaldson [79] established a fundamental mathematical result: there are four-dimensional manifolds that are homeomorphic (that is, topologically equivalent) to Euclidean space $\mathbb{R}_4$, but not diffeomorphic (that is, not equivalent in differential-geometric respect) to it, each of them being a smooth manifold. Such manifolds are called *exotic*. Later it was found that there is a whole continuum of four-dimensional exotic manifolds. In any other dimensions D there are no exotic smooth structures on $\mathbb{R}_D$. In other words, if $D \neq 4$, then any smooth manifold homeomorphic to $\mathbb{R}_D$ turns out to be automatically diffeomorphic to $\mathbb{R}_D$. The books [80] and [81] can serve as an introduction to the theory of exotic spaces.

The totality of exotic manifolds precisely forms an ensemble of rooms for quantum dramaturgy [21]. Note that objective mathematical grounds for combining spacetimes into an ensemble exist only for $D = 4$. If $D \neq 4$, then we have a set of clones of the Universe suitable for *imitation* of randomness, rather than the desired ensemble of *random* smooth structures of the Universe. Thus, the fundamentals of quantum physics, as we understand them today, are not statistically feasible in worlds with $D \neq 4$.

Exotic manifolds have not yet found their practical application in quantum theory [22]. To progress in this field is difficult even for top-notch mathematicians. The situation here is reminiscent of that developed at the beginning of the 17th century before the invention of rectangular coordinates and analytical geometry in the René Descartes magnum opus *Discours de la methode*.

5. Understanding Gravity

In an effort to derive the simplest expression for the gravitational four-force, with due regard to the phenomenology (the identity of inertial and gravitational masses and the deviation of a light path from a straight line under the influence of gravitational field of the Sun), we concluded in subsection 2.4. that a description of gravity is impossible without resorting to the concept of *spacetime warping*. Einstein was the first to come to this conclusion [3]. However, Einstein proceeded from other premises: the principle of equivalence and the condition that the dynamical laws be invariant under general coordinate transformations. If some result is obtained in two various ways, this gives an additional confidence of its correctness. This is all the more important in the present case because the consequences of the result obtained are extremely dramatic. When Minkowski spacetime $\mathbb{R}_{1,3}$ is abandoned in favor of pseudo-Riemannian manifolds $\mathfrak{R}_{1,3}$, the great scientific achievements

[21] One may prefer to speak about the quantum vacuum whose fluctuations give rise to four-dimensional rooms with diverse topologies and various exotic smooth structures.

[22] A possible way for exotic manifolds to appear in the formalism of quantum field theory is that the functional integral performs summation not only over all topological types of four-dimensional manifolds but also over all exotic smooth structures.

of the 18th and 19th centuries, the laws of conservation of energy-momentum and angular momentum, are sacrificed.

Emmy Noether proved a theorem [82], by which energy-momentum conservation is associated with *homogeneity* of spacetime, and angular momentum conservation is due to *isotropy* of space. In contrast to flat spacetime $\mathbb{R}_{1,3}$, curved manifolds $\mathfrak{R}_{1,3}$ do not, in general, possess these symmetry properties.

Poincaré believed that the choice of geometry is a matter of convention. There is no true geometry of our physical world. It is chosen for reasons of convenience. For Poincaré, Euclidean geometry of space $\mathbb{E}_3$ seemed the most convenient. Of course, one may adopt a different model of spacetime, and by appropriately changing physical laws discover that the entire set of physical events and connections between them remains intact [83]. This is a kind of principle of complementarity between geometry and physics.

This idea gave rise to attempts to construct theories of gravity in Minkowski space, with a symmetric tensor of the second rank $\phi_{\mu\nu}$ acting as the gravitational field [84]. In such theories, the effects of the general theory of relativity were reproduced for a *weak* gravitational field, that is, when all the components of $\phi_{\mu\nu}$ in $g_{\mu\nu} = \eta_{\mu\nu}+\phi_{\mu\nu}$ are small in any inertial frame of reference, $|\phi_{\mu\nu}| \ll 1$.

One further long-term project was to construct a bimetric theory of gravity with two arenas, $\mathfrak{R}_{1,3}$ and $\mathbb{R}_{1,3}$, where the dynamical affair in $\mathfrak{R}_{1,3}$ would be equivalent to that of general relativity [85]. The projection of $\mathfrak{R}_{1,3}$ onto $\mathbb{R}_{1,3}$ is to be translated into a theory of the gravitational field as an ordinary gauge field theory in which the energy-momentum and the angular momentum conservation laws are fulfilled. However, the equivalence of the dynamics is reachable only when the *topology* of $\mathfrak{R}_{1,3}$ is identical to that of $\mathbb{R}_{1,3}$, that is, again for weak fields. Attempts to invent a mechanism for automatically excluding solutions with non-trivial topology, such as solutions devoid of black holes, have fared poorly.

Much effort has been made to save conservation laws, mainly in relation to the 'island of matter surrounded by emptiness' [22]. In such models with any structure of the 'island', space is *asymptotically flat*, and this serves as a prerequisite for the Hamiltonian treatment of such a system [86].

The notion of an asymptotically flat space has no precise definition. Qualitatively, it is about the disappearance of curvature at infinity: $R^{\alpha}{}_{\beta\gamma\delta} \to 0$ as $r \to \infty$, and the choice of coordinates ensuring the convergence of additive quantities such as the Hamiltonian. However, these requirements are not restrictive enough to eliminate arbitrariness in foliating spacetime into spacelike sections consistent with asymptotic symmetries of the manifold $\mathfrak{R}_{1,3}$. Freedom in choosing the grid of spatial coordinates leads to the fact that the energy of a Schwarzschild black hole resulting from a collapsing star of mass M can take on any value greater than or equal to M [87].

The presence of ill-defined additive quantities in general relativity closely resembles the Banach–Tarski theorem [88] (the book [89] is devoted to a discussion of this theorem). By this theorem, a three-dimensional ball can be divided into several parts, which are then assembled by a continuous movement without changing their shape and without one part passing through another, to yield a ball twice as large as the original. The volume of the ball in the Banach–Tarski theorem is given by the usual three-dimensional Lebesgue measure, and the energy of a black hole is given by a linear functional whose measure is defined by the Arnowitt–Deser–Misner Hamiltonian [86].

It is not customary to discuss the Banach–Tarski paradox in the physics context for a simple reason: material bodies are made of atoms. The partitioning procedure of a ball is unrelated to their actual disintegration, and hence it is impossible to cut up a pea into several pieces and then reassemble them to form a Sun-sized ball. However, this objection overlooks one important instance, black holes. Each isolated stationary black hole is completely specified by three parameters: its mass M, angular momentum J and electric charge Q. All individual features of the collapsing system disappear in the black hole state. The event horizon of the black hole is stripped of the grain structure that was inherent in the collapsing system. The outer part of black holes is given by the Kerr–Newman solution. Unlike conventional quantum-mechanical bound systems, a black hole does not have a discrete energy spectrum. The linear functional associated with the energy of a black hole is foliation-dependent [87]. That is why this quantity allows the metamorphosis in the spirit of the Banach–Tarski theorem.

Gravity is not an ordinary physical field with uniquely defined energy-momentum. This makes it radically different from the other three fundamental interactions. It may be that all the energy inconsistencies, puzzles, and imbroglios related to gravity have a single origin. Planck considered the rationale behind the conservation of momentum to be the implementation of the action–reaction principle [90]. However, we have seen in subsection 3.5. that this principle has nothing to do with the general theory of relativity. It is reasonable to assume that the action–reaction principle and the energy-momentum conservation hold or fail together [45]. The gap in the description of gravity and the other three fundamental forces challenges the program of their unification—the cornerstone of modern physics.

6. Magic $\mathbb{R}_{1,3}$

Are we not falling into self-deception by attributing the properties $\mathbb{R}_{1,3}$ to reality? For more than two millennia people thought that Euclidean geometry is the only true reflection of the properties of space, until Nikolaĭ Ivanovich Lobachevskiĭ, Janos Bolyai, and Carl Friedrich Gauß shook this belief at the beginning of the

19th century. It transpired that there are infinitely many conceivable versions of consistent geometries.

Klein, having discovered the connection between the structure of a geometry and the group of allowable transformations of its elements, the automorphism group, was the first to realize that the faces of all (elementary) geometries can be completely classified by their automorphism groups [91]. The further development of mathematics brought to the forefront manifolds, curved geometric objects of arbitrary dimensions. However, the Klein approach still persists to be at a center of the layout, because a particular manifold is locally structured as a vector space whose metric form is defined via its finitely parametric automorphism group.

From the cognitive point of view, this circumstance is extremely favorable. It follows that although there are infinitely many geometries, their set can be captured by the eye and subjected to a clear classification. If we were lucky enough, we would get to the 'physically cherished' version of geometry. And indeed it happened.

Continuous simple groups were classified by Wilhelm Carl Joseph Killing, Elie Joseph Cartan, and Hermann Weyl. There are four series of classical groups: special unitary groups $\mathrm{SU}(n)$, special orthogonal groups of odd order $\mathrm{SO}(2n+1)$, symplectic groups $\mathrm{Sp}(2n)$, special orthogonal groups of even order $\mathrm{SO}(2n)$, together with five exceptional groups: G_2, F_4, E_6, E_7, E_8.

Among the listed groups there is a unique *special* case. The group SO(4) *is not simple*; it is a direct product of two simple groups, $\mathrm{SO}(3)\times\mathrm{SO}(3)$. Its complexification $\mathrm{SL}(4,\mathbb{C})$ has several real realizations. All of them are also not simple groups, with the only exception being the Lorentz group $\mathrm{SO}\,(1,3)$, see, *e. g.*, [95]. Now, with reference to Klein's Erlangen program [91], we take $\mathrm{SO}\,(1,3)$ as the automorphism group to arrive at the clearly distinguished geometry, the Minkowski spacetime geometry.

As noted in subsection 2.5., the principal characteristic trait of $\mathbb{R}_{1,3}$ is the light cone structure. Geometric constructions, defined by other automorphism groups—all, except $\mathrm{SO}\,(1,\mathrm{n})$—are deprived of such property. This, of course, does not mean they are physically useless. Indeed, automorphisms of phase spaces are given by symplectic groups $\mathrm{Sp}(2n)$, and automorphisms of the Hilbert state space of quantum mechanics are given by unitary groups. The absence of a light cone structure is not considered a flaw in these spaces. Only $\mathbb{R}_{1,n}$ are associated with the geometry of space and time. For example, string theories prove to be consistent only in manifolds locally identical to $\mathbb{R}_{1,9}$ or $\mathbb{R}_{1,23}$ [96], and an appropriate arena for M-theory are taken to be manifolds locally identical to $\mathbb{R}_{1,10}$ [97].

With the advent of exotic spaces, $\mathbb{R}_{1,3}$ was highlighted as the most perfect geometric model of reality. This discovery evidences that $D=4$ is indeed *objectively* distinguished.

In 1971, physicists of the USSR and the West invented *supersymmetry*, the symmetry group between fermionic and bosonic degrees of freedom [98]–[102]. Geometrically, such supergroups intertwine spacetime coordinates with Grassmannian variables attributable to field values. In supersymmetric models, ultraviolet divergences cancel in the one-loop approximation. Models showing ultraviolet finiteness in all orders of perturbation theory have also been found. The hope arose to construct a finite theory of all four fundamental forces, that is, to go one step further than Minkowski in revealing the geometric secrets of nature. However, no trace of supersymmetry has been found at the currently available collision energies of 13 TeV. Therefore, the idea of superspaces still remains an elegant game of the mind with unclear relationship to physical reality.

What is the future fate of $\mathbb{R}_{1,3}$? Predicting the future is said to be a risky business. Nevertheless, it seems worth taking a risk to suggest that we are on the threshold of new grandiose events in theoretical physics, probably related to the development of a simple technique for analyzing the quantum picture in terms of exotic spaces, and therefore a new look at $\mathbb{R}_{1,3}$.

References

[1] Minkowski, H. (1908). Raum und Zeit. *Phys. Z.* **10**, 75-88.

[2] Pais, A. (1982) '*Subtle is the Lord...' The Science and the Life of Albert Einstein* (Oxford: OUP), p. 152.

[3] Einstein, A., M. Grossmann (1913). Entwurf einer verallgemeinerten Relativitätstheorie und Theorie der Gravitation *Z. Math. Phys.* **62**, 225-261.

[4] Klein, F. (1897). *The Mathematical Theory of the Top.* (New York: Scribners).

[5] Poincaré, H. (1904). L'etat actuel et l'avenir de la Physique mathematiqie. *Bull. Sci. Math.* **28**, 302.

[6] Einstein, A. (1905). Zur Elektrodynamik der bewegter Körper. *Ann. Phys.* (Berlin) **322**, 891-921.

[7] Poincaré, H. (1906). Sur la dynamique de l'électron. *Rend. Circ. Mat. Palermo*, **21**, 129-176.

[8] Leibniz, G. W. (1710). *Essais de Théodicée sur la bonté de Dieu, la liberté de l'homme et l'origine du mal* (Amsterdam: Boudestein).

[9] Kosyakov, B. P. (2014). The pedagogical value of the four-dimensional picture I: Relativistic mechanics of point particles. *Eur. J. Phys.* **35**: 025012. ArXiv: physics/2205.01186.

[10] Kosyakov, B. P. (2014). The pedagogical value of the four-dimensional picture II: Another way of looking at the electromagnetic field. *Eur. J. Phys.* **35**: 025013. ArXiv: physics/2205.01181.

[11] Chubykalo, A. E., A. Espinoza, and B. P. Kosyakov (2016). The pedagogical value of the four-dimensional picture III: Solutions to Maxwell's equations. *Eur. J. Phys.* **37**: 045202. ArXiv: physics/2205.01326.

[12] Kosyakov, B. P. (2003). On the inert properties of particles in classical theory. *Particles & Nuclei*, **34**, 808-828. [*Fizika Ĕlementarnykh Chastits i Atomnogo Yadra*, **34**, 1563-1608]. ArXiv: hep-th/0208035.

[13] Landau, L. D. and E. M. Lifshitz. *The Classical Theory of Fields* (New York: Butterworth–Heinemann, 1980). Translated from the 6th Russian edition, 1973.

[14] The Berkeley physics course (1965). *Mechanics*, V. 1; *Electricity and Magnetism*, V. 2 (New York: McGraw-Hill).

[15] Kosyakov, B. P. (2008). Massless interacting particles. *J. Phys.* A **41**: 465401. ArXiv: hep-th/0705.1228.

[16] Seifert, H., W. Threlfall (1934). *Lehrbuch der Topologie* (Leipzig: Teubner).

[17] Planck, M. (1906). Das Prinzip der Relativität und die Grundgleichungen der Mechanik. *Verh. Dtsch. Phys. Ges.* **8**, 136-141.

[18] Hartle, J. B. (2003). *Gravitaty. An Introduction to Einstein's General Relativity* (San Francisco: Addison–Wesley).

[19] Jackson, J. D. (1962). *Classical Electrodynamics*. 2nd ed. 1975, 3rd ed. 1999. (New York: Willey).

[20] Wong, S. K. (1970). Field and particle equation for the classical Yang–Mills field and particles with isotopic spin. *Nuovo Cimento*, **A 65**, 689–694.

[21] Yang, C. N. and R. Mills (1954). Conservation of isotopic spin and isotopic gauge invariance. *Phys. Rev.* **96**, 191-195.

[22] Misner, C. W., K. S. Thorne, and J. A. Wheeler (1973). *Gravitation* (San Francisco: Freeman).

[23] Nordström, G. (1913). Träge und schwere Masse in der Relativitätsmechanik. *Ann. Phys.* (Berlin) **40**, 856-878; Zur Theorie der Gravitation vom Standpunkt des Relativitätsprinzip. *Ibid.*, **42**, 533-554.

[24] Alexandrov, A. D. and V. V. Ovchinnikova (1953). Notes on the basics of the theory of relativity. *Bulletin of the Leningrad State University*. Math., Phys., Chem. **11**, 95-110 (In Russian). A. D. Alexandrov (1996). *Selected Works*. Part 1. Selected Scientific Papers (Boca Raton: CRC).

[25] Taub, A. H. (1948). Orbits of charged particles in constant fields. *Phys. Rev.* **73**, 786-798.

[26] Helmholtz, H. (1858). Über Integrale der hydrodynamische Gleihungen, welche den Wirbelbewegungen entsprechen. *J. Reine Angew. Math.* **55**, 25-55.

[27] Born, M. and L. Infeld (1934). Foundations of the new field theory. *Proc. R. Soc. Lond.* A **144**, 425-451.

[28] Fradkin, E. S. and A. A. Tseytlin (1985). Non-linear electrodynamics from quantized strings. *Phys. Lett.* B **163**, 123-130.

[29] Bateman, H. (1909, 1910). The conformal transformations of a space of four dimensions and their applications to geometrical optics. *Proc. London Math. Soc.* **7**, 70-89; The transformation of the electrodynamical equations. *Ibid.*, **8**, 223-264.

[30] Cunningham, E. (1909). The principle of relativity in electrodynamics and an extension thereof. *Proc. London Math. Soc.* **8**, 77-98.

[31] Ibragimov, N. H. (1968). Group properties of wave equations for particles of zero mass. *Dokl. Akad. Nauk SSSR*, **178**, 566-568 (In Russian).

[32] Heaviside, O. (1892). *Electrical Papers*. V. I, II. (London: Macmillan).

[33] Larmor, J. (1900). *Aether and Matter* (Cambridge: CUP).

[34] Rainich, G. Y. (1925). Electrodynamics in general relativity. *Trans. Amer. Math. Soc.*, **27**, 106-136.

[35] Białynicki-Birula, I. (1983). Nonlinear electrodynamics: Variations on a theme by Born and Infeld. In: *Quantum Theory of Particles and Fields*. Birthday volume dedicated to Jan Łopuszański. Ed. B. Jancewicz and J. Lukierski. (Singapore: World Scientific), pp. 31-48.

[36] Weyl, H. (1918). *Raum, Zeit, Materie* (Berlin: Springer).

[37] Bandos, I., K. Lechner, D. Sorokin, and P. K. Townsend (2020). A non-linear duality-invariant conformal extension of Maxwell's equations. *Phys. Rev.* D **102**: 121703. ArXiv: hep-th/2007.09092.

[38] Kosyakov, B. P. (2020). Nonlinear electrodynamics with the maximum allowable symmetries. *Phys. Lett.* B **810**: 135840. ArXiv: hep-th/2007.13878.

[39] Dirac, P. A. M. (1931). Quantised singularities in the electromagnetic field. *Proc. R. Soc. Lond.* A **133**, 60-72.

[40] Dirac, P. A. M. (1948). The theory of magnetic poles. *Phys. Rev.* 2nd Ser. **74**, 817-830.

[41] Wu, T. T. and C. N. Yang (1975). Concept of nonintegrable phase factors and global formulation of gauge fields. *Phys. Rev.* D **12**, 3845-3857.

[42] Coleman, S. (1983). The magnetic monopole fifty years later. In: *The Unity of the Fundamental Interactions*. Ed. A. Zichichi (London: Plenum), pp. 21-117.

[43] Poincaré, H. (1896). Remarques sur une expérience de M. Birkeland. *C. R. Acad. Sci.* **123**, 530-533. *Œuvres*. T. 10 (Paris: Gauthier–Villars, 1954), pp. 310-313.

[44] Goddard, P. and D. I. Olive (1978). Magnetic monopoles in gauge field theories. *Rep. Prog. Phys.* **41**, 1357-1437.

[45] Chubykalo, A. E., A. Espinoza, and B. P. Kosyakov (2017). The origin of the energy-momentum conservation law. *Ann. Phys.* (N. Y.) **384**, 85-104. ArXiv: hep-th/1704.08123.

[46] Abraham, M. (1905). *Theorie der Elektrizität.* Bd II. (Leipzig: Teubner).

[47] Lorentz, H. A. (1909). *The Theory of Electrons and its Applications to the Phenomena of Light and Radiant Heat* (Leipzig: Teubner).

[48] Dirac, P. A. M. (1938). Classical theory of radiating electron. *Proc. R. Soc. Lond.* A **167**, 148-169.

[49] B. P. Kosyakov (1992). Radiation in electrodynamics and Yang–Mills theory. *Sov. Phys. Usp.* **35**, 135-142. [*Uspekhi Fizitsheskikh Nauk*, **162**, 161-176, 1992].

[50] Kosyakov, B. P. (1999). Exact solutions in classical electrodynamics and Yang–Mills theory in even-dimensional space-times. *Theor. Math. Phys.* **119**, 493-505 (1999). [*Teoretitsheskaya i Matematitsheskaya Fizika*, **119**, 119-135, 1999]. ArXiv: hep-th/0207217.

[51] Kosyakov, B. P. (2008). Electromagnetic radiation in even-dimensional spacetimes. *Int. J. Mod. Phys.* A **23**, 4695-4708. ArXiv: hep-th/0803.3304.

[52] Kosyakov, B. P. (2019). Self-interaction in classical gauge theories and gravitation. *Phys. Rep.* **812**, 1-56. ArXiv: hep-th/1812.03290.

[53] Teitelboim, C. (1970). Splitting of Maxwell tensor: Radiation reaction without advanced fields. *Phys. Rev.* D **1**, 1572-1582.

[54] Larmor, J. (1897). On the theory of magnetic influence on spectra; and on the radiation from moving ions. *Philos. Mag.* **44**, 503-512.

[55] Heaviside, O. (1902). The waste of energy from a moving electron. *Nature*, **67**, 6-8.

[56] Kosyakov, B. P. (1998). Exact solutions in the Yang–Mills–Wong theory. *Phys. Rev.* D **57**, 5032-5048. ArXiv: hep-th/9902039.

[57] Planck, M. (1908). Bemerkungen zum Prinzip der Aktion Reaktion in der allgemeinen Dynamik. *Phys. Z.* **9**, 828-830.

[58] Kant, I. (1781). *Kritik der reinen Vernunft* (Riga: Hartknoch).

[59] Marx, K. (1903). Grundrisse der Kritik der politischen Ökonomie (Rohentwurf). *Neue Zeit* **1**, 23-35.

[60] Fokker, A. D. (1929). Ein invarianter Variationssatz für die Bewegung mehrerer elektrischer Massenteilchen. *Z. Phys.* **58**, 386-393.

[61] Wheeler, J. A. and R. P. Feynman (1945). Interaction with the absorber as the mechanism of radiation. *Rev. Mod. Phys.* **17**, 157-181 .

[62] Wheeler, J. A. and R. P. Feynman (1949). Classical electrodynamics in terms of direct interparticle action. *Rev. Mod. Phys.* **21**, 425-433.

[63] Hoyle, F. and J. V. Narlikar (1995). Cosmology and action-at-a-distance electrodynamics. *Rev. Mod. Phys.* **67**, 113-155.

[64] Ryder, L. H. (1974). Conformal invariance and action-at-a-distance electrodynamics. *J. Phys.* A **7**, 1817-1828.

[65] Boulware, D. G., L. S. Brown, and R. D. Peccei (1970). Deep-inelastic electroproduction and conformal symmetry. *Phys. Rev.* D **2**, 293-298.

[66] Pegg, D. T. (1975). Absorber theory of radiation. *Rep. Prog. Phys.* **38**, 1339-1383.

[67] Schweber, S. S. (1964). *An Introduction to Relativistic Quantum Field Theory* (New York: Row). Ch. 13.

[68] Feynman, R. P. (1966). The development of the space-time view of quantum mechanics. Nobel lecture. *Science*, **153**, 699-708.

[69] Feynman, R. P. and A. R. Hibbs (1965). *Quantum Mechanics and Path Integrals* (New York: McGraw-Hill).

[70] Brink, L., S. Deser, B. Zumino, P. Di Vecchia, P. Howe (1976). Local supersymmetry for spinning particles. *Phys. Lett.* B **64**, 435-438.

[71] Wick, G. C. (1954). Properties of Bethe–Salpeter wave functions. *Phys. Rev.* **96**, 1124-1134.

[72] Glimm, J. and A. Jaffe (1981). *Quantum Physics. A Functional Integral Point of View*. (Berlin: Springer).

[73] Kosyakov B. P. (2008). Black holes: interfacing the classical and the quantum. *Found. Phys.* **38**, 678-694. ArXiv: gr-qc/0707.2749.

[74] Streater, R. F., and A. S. Wightman (1964). *PCT, Spin and Statistics and All That* (New York: Benjamin).

[75] Bogolubov, N. N. , A. A. Logunov, A. I. Oksak, and I. T. Todorov (1990). *General Principles of Quantum Field Theory* (Dordrecht: Kluwer). (Russian original published in 1987).

[76] Blokhintsev, D. I. (1968). *The Philosiphy of Quantum Mechanics* (Dordrecht: Reidel). (Russian original published in 1966).

[77] Home D. and M. A. Whitaker (1992). Ensemble interpretation of quantum mechanics: A modern perspective. *Phys. Rep.* **210**, 225-317.

[78] Freedman, M. (1982). The topology of four-dimensional manifolds. *J. Diff. Geom.* **17**, 357-453.

[79] Donaldson, S. (1983). An application of gauge theory to four-dimensional topology. *J. Diff. Geom.* **18**, 279-315.

[80] Gompf R. E. and A. I. Stipsicz (1999). *4-Manifolds and Kirby Calculus* (Providence: AMS).

[81] Scorpan, A. (2005). *The Wild Word of 4-Manifolds* (Providence: AMS).

[82] Noether, E. (1918). Invariante Variationsprobleme. *Nachr. Ges. Wiss. Göttingen*. **2**, 235-258.

[83] H. Poincaré. (1902). *La science et l'hypothèse*. (Paris: Flammarion).

[84] Rosen, N. (1940). General relativity and flat space. I and II, *Phys. Rev.* **15**, 147-150, and 150-153.

[85] Logunov A. A. (2001). *The Theory of Gravity* (Moscow: Nauka) (Russian original published in 2000). ArXiv: gr-qc/0210005.

[86] Arnowitt, R., S. Deser, and C. W. Misner (1960). Canonical variables for general relativity. *Phys. Rev.* **117**, 1595-1602. Energy and the criteria for radiation in general relativity. *Ibid.* **118**, 1100-1104. Coordinate invariance and energy expressions in general relativity. *Ibid.* **122**, 997-1006.

[87] Denisov, V. I. and V. O. Solov'ev (1983). The energy determined in general relativity on the basis of the traditional Hamilton approach does not have physical meaning. *Theor. Math. Phys.* **56**, 832-841. [*Teoretitsheskaya i Matematitsheskaya Fizika*, **56**, 301-314, 1983].

[88] Banach, S. and A. Tarski (1924). Sur la décomposition des ensembles de points en parties respectivement congruentes. *Fund. Math.* **6**, 244-277.

[89] Wagon, S. (1994). *The Banach–Tatski Paradox* (Cambridge: CUP).

[90] Planck, M. (1908). Bemerkungen zum Prinzip der Aktion Reaktion in der allgemeinen Dynamik. *Phys. Z.* **9**, 828-830.

[91] Klein, F. (1893). Vergleichende Betrachtungen über neuere geometrische Forschungen. *Math. Ann.* **43**, 63-100; *Gesam. Abh.* Bd 1 (Springer: Berlin, 1921), S. 460-497.

[92] Killing, W. (1888-1890). Die Zusammensetzung der stetigen endlichen Transformationsgruppen. *Math. Ann.* **31**, 252-290; **33**, 1-48; 1889. **34**, 57-122; 1890; **36**, 161-189.

[93] Cartan, É. (1894). Sur la structure des groupes de transformations finis et continus. *Œuvres complètes*. Paris. T. 1, pp. 137-287.

[94] Weyl, H. (1935). *The Structure and Representations of Continuous Groups*. Lecture Notes (Princeton: PUP).

[95] Barut, A. O., and R. Rączka (1977). *Theory of Group Representations and Applications*. 2nd ed. 1980 (Warszawa: PWN–Polish Sci. Pub.).

[96] Green, M. B., J. H. Schwarz, and E. Witten (1987). *Superstring Theory*. V. 1, 2 (Cambridge: CUP).

[97] Polchinski, J. (1996). String duality. *Rev. Mod. Phys.* **68**, 1245-1258. ArXiv: hep-th/9607050.

[98] Berezin, F. A. and G. I. Kats (1970). Lie groups with commuting and anticommuting parameters. *Math. USSR-Sb.* **11**, 311-325. [*Matematitsheskiĭ Sbornik*, **82**, 343-359, 1970].

[99] Gol'fand, Yu. A. and E. P. Likhtman (1971). Extension of the algebra of Poincaré group generators and violation of P-invariance. *JETP Lett.* **13**, 323-326. [*Pis'ma v ZhETF*, **13**, 452-455, 1971].

[100] Neveu, A. and J. H. Schwarz (1971). Factorizable dual model of pions. *Nucl. Phys.* B **31**, 86-112.

[101] Gervais, J.-L. and B. Sakita (1971). Field theory interpretation of supergauges in dual models. *Nucl. Phys.*, B **34**, 632-639.

[102] Ramond, P. (1971). Dual theory for free fermions. *Phys. Rev.* D **3**, 2415-2418.

Chapter 5

Bivector Gauge Fields and Amplification of Radiation from Distant Cosmic Sources

Alexander Krasulin
Institute for Nuclear Research of the Russian Academy of Sciences*

Abstract

This chapter describes in some detail the concept of bivector gauge fields and then examines the possibility of confirming their existence by recording certain patterns in variation of intensity of radiation from distant cosmic sources versus radiation frequency and versus the estimated distance to the source.

Keywords: bivector gauge fields, five-vector tangent vectors

This chapter deals with the so-called *bivector* gauge fields; more precisely, with the kind that have to do with electromagnetism. The concept of bivector gauge fields arises rather naturally within the theory of five-dimensional tangent vectors in space-time, presented in ref.[2][1]. However, it is quite possible to explain what these gauge fields are in terms of ordinary four-vectors and four-tensors as well. Though in this case it will not be so clear why they have been named 'bivector'. Nor why anyone should have thought of a thing like that in the first place!

To explain what bivector gauge fields are one should recall what is parallel transport. As is known, the latter is a procedure by means of which one is able to transport a non-scalar quantity—say, a tangent vector or a spinor or any other vector-

---

*Former affiliation. Corresponding author's email: krasulin@post.com.

[1] A brief introduction can be found in ref.[2].

In: Mathematical Problems in Relativity, Gravitation, and Cosmology
Editor: Valeriy Dvoeglazov
ISBN: 979-8-89530-622-2

or tensor-like object that one can think of—from one space-time point to another. Such transportation is needed to calculate the derivative of non-scalar-valued fields. The simplest example of transport rules are those for ordinary tangent four-vectors. Since it is always assumed that parallel transport is a linear operation, the transport rules in this case can be expressed with a set of 64 scalar quantities called connection coefficients and typically denoted as ${\Gamma^\alpha}_{\beta\mu}$. The latter have the following meaning: An arbitrary four-vector with components V^α at a point with coordinates x^μ, when transported to a nearby point with coordinates $x^\mu + dx^\mu$, will have components equal to

$$V^\alpha - \Gamma^\alpha_{\beta\mu} V^\beta dx^\mu$$

plus quantities of higher order in dx^μ. Knowing the connection coefficients, one can evaluate the covariant derivative of an arbitrary four-vector field U^α: The latter will have the components

$$U^\alpha_{;\mu} = \partial_\mu U^\alpha + \Gamma^\alpha_{\beta\mu} U^\beta. \tag{1}$$

In physics, the rules of parallel transport for four-vectors are regarded as a special kind of geometry possessed by space-time, which is related to the Riemannian geometry fixed by the metric, but is not necessarily determined by the latter completely. The situation is that the metric, by itself, determines certain transport rules for four-vectors; it is these rules that one considers in General Relativity. However, in a given Riemannian geometry, the actual parallel transport rules—the ones that are used to calculate the derivatives of matter fields in field equations—may differ from the rules defined by the metric. The latter fact will manifest itself in that transported four-vectors will experience additional rotation compared to the same vectors but transported according to the rules fixed by the metric. This additional rotation can be described with the quantities $S^\alpha_{\beta\mu}$ equal to the difference between the connection coefficients corresponding to the transport fixed by the metric and the connection coefficients corresponding to the actual transport rules. With transformation of the four-vector basis, the quantities $S^\alpha_{\beta\mu}$ transform as components of a four-tensor, which is called *torsion*, or more precisely, the *contorsion* tensor.

Another important case of parallel transport is that corresponding to abstract vectors and tensors associated with internal symmetry groups in elementary particle physics. Such vectors and tensors are not related in their nature to the space-time manifold and in the following will be generally referred to as *nonspacetime* vectors and tensors. As in the case of tangent four-vectors, one can introduce the corresponding connection coefficients, which in this case are called gauge fields. This will enable one to calculate the covariant derivative of any field whose values are nonspacetime vectors or tensors of a given kind. If $G^i_{j\mu}$ are the gauge fields that correspond to the considered type of nonspacetime vectors, then the components of the covariant derivative of any field W^i whose values are such nonspacetime vectors are given by a formula similar to formula (1):

$$W^i_{\;;\mu} = \partial_\mu W^i + eG^i_{\;j\mu} W^j,$$

where, as is customary, we have written out explicitly the 'charge' e of the field W^i relative to the interaction mediated by the gauge fields $G^i_{\;j\mu}$.

It is practically always supposed implicitly that parallel transport of nonspacetime vectors is independent of torsion. The concept of bivector gauge fields offers a scheme where this is not so. To see how it works, one should introduce the notion of the *bivector* derivative. Let us first define the latter for four-vector fields. For simplicity, let us consider flat space-time and introduce in it an arbitrary system of Lorentz coordinates x^μ. The *covariant* derivative of an arbitrary four-vector field $\mathbf{U}$ corresponding to the transport rules fixed by the metric can then be evaluated by using the following four-step procedure: One performs an infinitesimal active *translation* of the differentiated field in the direction specified by the argument of the derivative; subtracts from the result the original field; divides the difference by the translation parameter s; and takes the limit $s \to 0$. Consider now a more general procedure where one takes the same field $\mathbf{U}$, but this time performs an *arbitrary* active Poincare transformation, which may include both translations *and* rotations. The rest of the procedure is the same as before: One subtracts from the result the original field; divides the difference by the transformation parameter s; and takes the limit $s \to 0$. As a result, one obtains a field which will be called the *bivector derivative* of field $\mathbf{U}$ and will be denoted as $\mathsf{D}_\mathcal{A}\mathbf{U}$, where $\mathcal{A}$ symbolizes the argument of derivative D. Let us determine what kind of quantity this argument is.

In terms of four-vectors, an infinitesimal Poincare transformation can be described with a pair of quantities $(\delta\mathbf{T}, \delta\mathbf{\Omega})$, where $\delta\mathbf{T}$ is a four-vector that describes the infinitesimal translation and $\delta\mathbf{\Omega}$ is an antisymmetric four-tensor of rank two that describes the infinitesimal four-dimensional rotation. As is known, such a description of Poincare transformation is *not* coordinate-independent, since the four-vector $\delta\mathbf{T}$ depends on the choice of the coordinate origin. At the same time, the active Poincare transformation itself is an operation that exists regardless of the choice of the coordinate system, so it may be expected that a coordinate-independent description for it does exist. This is where five-dimensional tangent vectors and the tensors constructed from them come in handy. It turns out that an infinitesimal Poincare transformation can be described in a coordinate-independent way with an antisymmetric *five*-tensor $\delta\mathcal{R}$ of rank two. The latter has the following components:

$$(\delta\mathcal{R})^{\mu\nu} = (\delta\mathbf{\Omega})^{\mu\nu}, \quad (\delta\mathcal{R})^{\mu 5} = -\,(\delta\mathcal{R})^{5\mu} = (\delta\mathbf{T})^\mu,$$

where μ and ν, as all Greek indices, run 0, 1, 2, and 3, and the index value 5 corresponds to the additional fifth component (for more details the reader is referred to ref.[1]). Dividing $\delta\mathcal{R}$ by s and taking the limit $s \to 0$, one obtains a finite 'five-dimensional' bivector $\mathcal{A}$, which uniquely determines the corresponding infinitesimal Poincare transformation to the first order, and therefore can serve as the argument of derivative D. This explains why the latter has been named the *bivector* derivative.

From the definition of the bivector derivative presented above, it is easy to obtain the following formulae for the components of D of an arbitrary four-vector field $\mathbf{U}$ in a Lorentz four-vector basis in flat space-time:

$$\begin{aligned}(\mathsf{D}_{\mu 5}\mathbf{U})^\alpha &= -(\mathsf{D}_{5\mu}\mathbf{U})^\alpha = \partial_\mu U^\alpha \\ (\mathsf{D}_{\mu\nu}\mathbf{U})^\alpha &= -(\mathsf{D}_{\nu\mu}\mathbf{U})^\alpha = (M_{\mu\nu})^\alpha_{\ \beta}\, U^\beta,\end{aligned}$$

where $\mathsf{D}_{\mu 5}$ denotes the bivector derivative that corresponds to translation along the axis x^μ, $\mathsf{D}_{\mu\nu}$ denotes the bivector derivative corresponding to rotation in the plane $x^\mu x^\nu$, and $(M_{\mu\nu})^\alpha_{\ \beta} \equiv \delta^\alpha_{\ \nu}\, g_{\mu\beta} - \delta^\alpha_{\ \mu}\, g_{\nu\beta}$. These formulae will also be valid in local Lorentz coordinates in space-time with arbitrary curvature. Comparing them with the formula for the components of the covariant derivative of field $\mathbf{U}$ in (local) Lorentz coordinates in space-time with arbitrary torsion:

$$\begin{aligned}(\nabla_\mu \mathbf{U})^\alpha &= \partial_\mu U^\alpha + \Gamma^\alpha_{\ \beta\mu} U^\beta = \partial_\mu U^\alpha - S^\alpha_{\ \beta\mu} U^\beta \\ &= \partial_\mu U^\alpha - S^{\sigma\tau}_{\ \ \ \mu}\, \delta^\alpha_{\ \sigma} g_{\tau\beta}\, U^\beta \\ &= \partial_\mu U^\alpha + \tfrac{1}{2} S^{\sigma\tau}_{\ \ \ \mu} (M_{\sigma\tau})^\alpha_{\ \beta} U^\beta,\end{aligned}$$

one observes that derivatives D and ∇ of field $\mathbf{U}$ are related to each other in the following way:

$$\nabla_\mu \mathbf{U} = \mathsf{D}_{\mu 5}\mathbf{U} + \tfrac{1}{2}\, S^{\sigma\tau}_{\ \ \ \mu}\, \mathsf{D}_{\sigma\tau}\mathbf{U}. \tag{2}$$

In the case of four-vector fields, the latter equation is a *consequence* of the definition of derivatives D and ∇. To define the bivector derivative for a field whose values are nonspacetime vectors or tensors of some particular kind, one *postulates* that (*a*) for such a field the derivative D exists and that (*b*) it is related to derivative ∇ as in equation (2).

As in the case of the covariant derivative, one can introduce the analogs of connection coefficients for D. For the components of the bivector derivative of an arbitrary field W^i one will then have:

$$\begin{aligned}(\mathsf{D}_{\mu 5} W)^i &= \partial_\mu W^i + e C^i_{\ j\mu 5} W^j \\ (\mathsf{D}_{\mu\nu} W)^i &= e C^i_{\ j\mu\nu} W^j.\end{aligned} \tag{3}$$

The fields $C^i_{j\mu 5}$ and $C^i_{j\mu\nu}$ are called *bivector gauge fields*. With transformation in the space of corresponding nonspacetime vectors, say, $W^i \mapsto W'^i = L^i_j W^j$, the bivectors gauge fields transform as follows:

$$\begin{aligned} C'^i_{j\mu 5} &= L^i_k C^k_{l\mu 5}(L^{-1})^l_j + e^{-1}L^i_k \partial_\mu (L^{-1})^k_j \\ C'^i_{j\mu\nu} &= L^i_k C^k_{l\mu\nu}(L^{-1})^l_j. \end{aligned} \tag{4}$$

So the fields $C^i_{j\mu 5}$ transform as ordinary gauge fields, while the fields $C^i_{j\mu\nu}$ transform as components of a tensor of rank (1,1) over the space of corresponding nonspacetime vectors.

If one assumes that the Lagrangian density for matter fields in local field theory depends on the value of the field at a given point and on the value of its *covariant* derivative, then the considered scheme implies that in the absence of torsion the matter fields are affected only by the gauge fields $C^i_{j\mu 5}$, and it is the latter that play the role of 'ordinary' gauge fields in this case: $G^i_{j\mu} = C^i_{j\mu 5}$. Under such circumstances, the fields $C^i_{j\mu\nu}$ have no direct effect on matter fields regardless of whether they are nonzero or not. At nonzero torsion, matter fields 'feel' both types of bivector gauge fields, and one has

$$G^i_{j\mu} = C^i_{j\mu 5} + \tfrac{1}{2} C^i_{j\sigma\tau} S^{\sigma\tau}{}_{\mu}. \tag{5}$$

Figuratively speaking, the fields $C^i_{j\mu\nu}$ *translate* the additional rotation of tangent four-vectors, representing torsion, into additional rotation of the relevant nonspacetime vectors. One can also see that the affine geometry represented by the gauge fields $C^i_{j\mu\nu}$ is, so to say, *hidden* in the sense that it has a noticeable direct effect on matter fields only when the torsion is sufficiently large. Fortunately, as will be shown below, the existence of this type of bivector gauge fields should also manifest itself *indirectly*: in that the fields $C^i_{j\mu 5}$—whose effect on matter fields is not suppressed by torsion, and which have all the properties of conventional gauge fields—*created* by matter fields will *differ* from the gauge fields these same matter fields would have created within the traditional gauge field theory framework.

The rest of this chapter is devoted to the bivector gauge fields associated with electromagnetism. In this case the fields $C_{\mu 5}$ and $C_{\mu\nu}$ will apparently have only two lower space-time indices and no indices associated with internal degrees of freedom. In order to construct the Lagrangian density that would determine the dynamics of such fields, one should first introduce the analog of the field strength tensor for the derivative D. As is shown in part VI of ref.[1], the latter is a 'five-dimensional' tensor F whose components have four lower indices and the following symmetry properties:

$$\mathsf{F}_{ABCD} = -\,\mathsf{F}_{BACD} = -\,\mathsf{F}_{ABDC}$$

$$\mathsf{F}_{ABCD} = -\,\mathsf{F}_{CDAB},$$

where indices A, B, C, and D run 0, 1, 2, 3, and 5. The components of F are the following:

$$\mathsf{F}_{\mu 5\alpha 5} = iF_{\mu\alpha}, \ \mathsf{F}_{\mu 5\alpha\beta} = -\mathsf{F}_{\alpha\beta\mu 5} = \partial_\mu C_{\alpha\beta},$$
$$\mathsf{F}_{\mu\nu\alpha\beta} = -\ g_{\mu\alpha}C_{\nu\beta} + g_{\nu\alpha}C_{\mu\beta} + \ g_{\mu\beta}C_{\nu\alpha} - g_{\nu\beta}C_{\mu\alpha},$$

where we have introduced the notation $A_\alpha = -iC_{\alpha 5}$ and where $F_{\alpha\beta} = \partial_\alpha A_\beta - \partial_\beta A_\alpha$. One may then suppose that, similar to the case of ordinary gauge fields, the Lagrangian density for their bivector analogs should be some bilinear combination of the components of F. Observing now that, from the latter, one can construct only *two* independent true scalars, for example,

$$\mathsf{F}^{ABCD}\mathsf{F}_{ABCD} \ \text{ and } \ \mathsf{F}^{AC}{}_{AD}\,\mathsf{F}^{BD}{}_{BC},$$

one obtains the following general expression for the Lagrangian density:

$$a \cdot \mathsf{F}^{ABCD}\mathsf{F}_{ABCD} + b \cdot \mathsf{F}^{AC}{}_{AD}\,\mathsf{F}^{BD}{}_{BC}\ , \tag{6}$$

where a and b are some unknown coefficients. Determining the value of the latter from the requirement that combination (6) reproduces the standard kinetic term for the field A_α (which can be regarded as ordinary electromagnetic potential) and the kinetic term for the antisymmetric tensor field (see e.g. ref.[3]), one arrives at the following Lagrangian density:

$$\begin{aligned} -\tfrac{1}{4}F^{\alpha\beta}F_{\alpha\beta} + \tfrac{1}{4}\,\partial^\mu K^{\alpha\beta}\partial_\mu K_{\alpha\beta} \\ -(\partial^\mu K_{\mu\alpha})^2 + \ 2\kappa\, F^{\alpha\beta}K_{\alpha\beta} - \tfrac{3}{2}\kappa^2\, K^{\alpha\beta}K_{\alpha\beta}, \end{aligned} \tag{7}$$

where $K_{\alpha\beta} = -i\kappa C_{\alpha\beta}$. Here κ is a certain universal constant with dimension of inverse length that appears within the theory of five-dimensional tangent vectors, and whose value is not fixed by mathematics (see ref.[1]). As one can see, in addition to the kinetic terms for A_α and $K_{\alpha\beta}$, one obtains a term where A_α and $K_{\alpha\beta}$ mix, and a term that has the form of a mass term for the field $K_{\alpha\beta}$. Owing to the first of these additional terms, the equations for the electromagnetic field and for the field $K_{\alpha\beta}$ are no longer independent from each other and have the following form:

$$\partial^\alpha F_{\alpha\beta} = 4\kappa\,\partial^\alpha K_{\alpha\beta} + j_\beta, \tag{8}$$

$$\partial^2 K_{\alpha\beta} + 2\,\partial^\lambda(\partial_\alpha K_{\beta\lambda} - \partial_\beta K_{\alpha\lambda}) + 6\kappa^2 K_{\alpha\beta} = 4\kappa F_{\alpha\beta} + j_{\alpha\beta}, \tag{9}$$

where j_α and $j_{\alpha\beta}$ are obtained by varying the part of the action that describes the interaction of matter fields with bivector gauge fields with respect to A_α and $K_{\alpha\beta}$, respectively. It is useful to observe that, under the assumption made above about Lagrangian density depending on the *covariant* derivative of matter fields, one has

$$j_{\alpha\beta} = \kappa^{-1} S_{\alpha\beta}{}^{\mu}\, j_\mu,$$

where $S_{\alpha\beta}{}^{\mu} = g_{\alpha\sigma}\, g_{\beta\tau}\, S^{\sigma\tau}{}_{\omega}\, g^{\omega\mu}$. So the components $j_{\alpha\beta}$ will be comparable with j_α only if contorsion is of the order of constant κ.[2]

As mentioned above, apart from its direct action on matter fields (which is suppressed by torsion), the field $K_{\alpha\beta}$ manifests its existence in that the electromagnetic potential A_α created by some set of sources will *differ* from the potential these same sources would have created according to conventional electrodynamics. To see that this is indeed so, let us derive, from the general equations (8) and (9), the equations that specifically determine the electromagnetic potential. To this end, let us calculate the 4-divergence of both sides of equation (9) with respect to index α. Multiplying both sides of the equation by $\frac{2}{5}\kappa^{-1}$ and introducing the notation $C_\beta \equiv \frac{2}{5}\kappa^{-1}\partial^\alpha K_{\alpha\beta}$, one obtains

$$-\partial^2 C_\beta + 6\kappa^2 C_\beta = \tfrac{8}{5}\,\partial^\alpha F_{\alpha\beta} + \tfrac{2}{5}\kappa^{-1}\partial^\alpha j_{\alpha\beta}.$$

Substituting the right-hand side of equation (8) instead of $\partial^\alpha F_{\alpha\beta}$, and rearranging the terms, one gets

$$\partial^2 C_\beta + 10\kappa^2 C_\beta = -\tfrac{8}{5}\,t_\beta - \tfrac{8}{5}\,j_\beta, \tag{10}$$

where $t_\beta \equiv \frac{1}{4}\kappa^{-1}\partial^\alpha j_{\alpha\beta}$. If the electromagnetic potential obeys the Lorentz condition $\partial^\alpha A_\alpha = 0$, equation (8) acquires the form

$$\partial^2 A_\beta = 10\kappa^2 C_\beta + j_\beta. \tag{11}$$

Expressing $10\kappa^2 C_\beta$ in terms of j_β, t_β, and $\partial^2 C_\beta$ via equation (10), one can present equation (11) as

$$\partial^2 (A_\beta + C_\beta) = (1 - \tfrac{8}{5}) j_\beta - \tfrac{8}{5} t_\beta.$$

Finally, introducing the notation $B_\beta \equiv A_\beta + C_\beta$, one arrives at the following equations:

$$\partial^2 B_\beta = -\tfrac{8}{5} t_\beta - \tfrac{8}{5} j_\beta + j_\beta \tag{12a}$$

$$10\kappa^2 C_\beta + \partial^2 C_\beta = -\tfrac{8}{5}\,t_\beta - \tfrac{8}{5}\,j_\beta \tag{12b}$$

$$A_\beta = B_\beta - C_\beta. \tag{12c}$$

[2] Another possibility is that for some matter fields the Lagrangian density would depend on their bivector derivative *directly*, not via ∇ (see e.g. ref.[4]).

As one can see, if the field C_β were massless, too, the contributions to B_β and C_β created by the source $-\frac{8}{5}t_\beta - \frac{8}{5}j_\beta$ would be exactly the same and would cancel out each other in the potential A_β. So the latter would be exactly what it should be according to conventional electrodynamics. To see how the nonzero mass of C_β changes the potential, let us consider the solution of equations (12) for a static point source at the coordinate origin. As usual, 'static' and 'point' mean that $j_i = 0$ and $j_0 = e\delta(\vec{r})$ and that all the time derivatives are zero. For simplicity, we will also suppose that torsion is so small that the contribution of t_α to the electromagnetic potential is insignificant. The solution is the following:

$$B_i = C_i = 0, \quad B_0 = \left(1 - \tfrac{8}{5}\right)\tfrac{e}{4\pi r}, \quad C_0 = -\tfrac{8}{5}\tfrac{e}{4\pi r}e^{-\mu r},$$

where $\mu = \sqrt{10}\kappa$. Thus, for the electromagnetic potential one has:

$$A_i = 0 \text{ and } A_0 = \tfrac{e}{4\pi r} - \tfrac{8}{5}\tfrac{e}{4\pi r}(1 - e^{-\mu r}). \tag{13}$$

It is apparent that at $r \ll \mu^{-1}$, the second term in the expression for A_0 is practically zero, and the scalar potential itself is given by the usual formula: $A_0 = (e/4\pi r)$. As r increases, the second term in the expression for A_0 grows, and at $r = \ln(\frac{8}{3})\,\mu^{-1} = 0.9808\,\mu^{-1}$ the potential turns to zero and then changes its sign, as is shown in Figure 1. Here the potential is measured in the units of

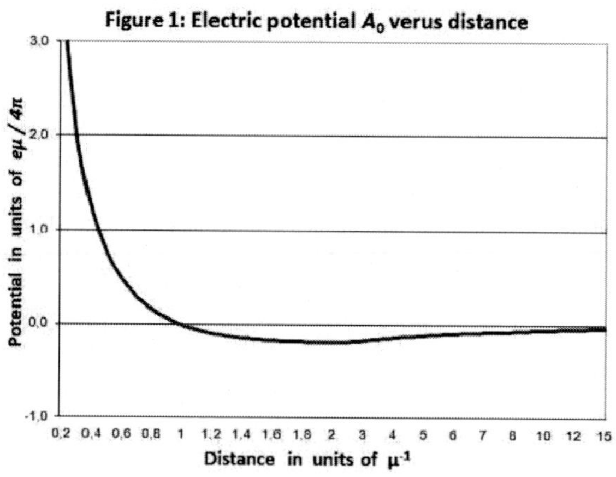

$e\mu/4\pi$ and r is measured in the units of μ^{-1}. With further increase of r, one observes a 'valley', where the potential changes very little and after reaching a minimum, begins to grow again. This means that within this distance range, the electric field decreases and turns to zero and then changes to the opposite direction. At $r \gg \mu^{-1}$, one has $A_0 = -\frac{3}{5}\cdot(e/4\pi r)$, as if this was a Coulomb potential created by the charge $-\frac{3}{5}e$. As in conventional electrodynamics, the corresponding electric field will have only the radial component, which can be presented in the following form:

$$E_r = \tfrac{e\mu^2}{4\pi}\mathcal{E}(\mu r),$$

where

$$\mathcal{E}(z) = \tfrac{8}{5}\tfrac{e^{-z}}{z^2}\,(1+z) - \tfrac{3}{5}\tfrac{1}{z^2}.$$

As one can see, at distances much less than μ^{-1}, the interaction between two charged particles obeys the Coulomb law. The behavior of the electric field at r greater than μ^{-1} is shown in Figure 2. As the distance between the particles in-

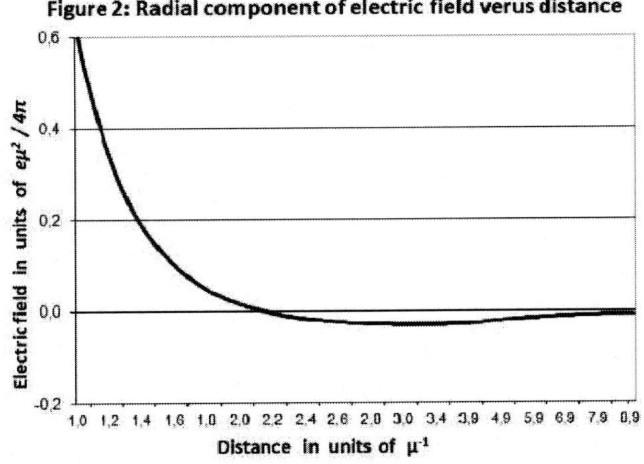

Figure 2: Radial component of electric field verus distance

creases and approaches a critical value of $r_c = 2.118\,\mu^{-1}$, the magnitude of the force acting between them rapidly diminishes and turns to zero at $r = r_c$. With further increase of the distance, the electrostatic interaction between the particles appears again, but now it is of opposite direction: charges of the same sign now *attract* and charges of opposite signs *repel*. With further increase of r, the interaction between the particles continues to grow until it reaches a maximum around $r = 3.224\,\mu^{-1}$, after which its magnitude begins to decrease. At $r \gg \mu^{-1}$, the force of interaction between the particles is again inversely proportional to r^2, only now its magnitude is smaller than the Coulomb interaction in conventional electrostatics by a factor of $\frac{3}{5}$, and one has attraction instead of repulsion and vice versa.

From all these observations one can conclude that constant κ should be either very small or very large, so that the deviation of potential (13) from the Coulomb form would lie outside the range where the Coulomb law has been tested. If κ were *large*, and accordingly, the distance μ^{-1} were small, the described unorthodox behavior of the electric field would be a *quantum* effect and should be examined within quantum field theory. However, at macroscopic distances attraction of charges of the same sign and repulsion of charges of opposite signs should then have been observed, which is not what is seen in reality.

One is therefore left with the possibility that constant κ is *small*, and accordingly, the distance μ^{-1} is large. In this case, the considered deviation from the Coulomb law is a *macroscopic* effect and is actually a deviation from classical electrodynamics. It is obvious that measuring the deviation from the Coulomb law at large distances directly is not an option. A more realistic way to confirm the existence of bivector gauge fields, and to measure the constant κ, is to examine monochromatic radiation from far away sources. To simplify the calculations, let us again suppose that the contribution of t_β is insignificant. Then, according to equations (12), at large distances from the source, the radiation would approximately be a superposition of two monochromatic plane waves: one with the wavelength $\lambda_1 = 2\pi c\omega^{-1}$ and relative amplitude $-\frac{3}{5} = 1 - \frac{8}{5}$, the other with the wavelength $\lambda_2 = 2\pi(\omega^2 c^{-2} - \mu^2)^{-1/2}$ and relative amplitude $+\frac{8}{5}$, where ω is the common angular frequency of both waves. Let us suppose that the source is located at the coordinate origin. If μ^{-1} is a very large macroscopic distance, then for all the electromagnetic radiation recorded from the source one evidently has $\mu c/\omega \ll 1$. This means that one can safely disregard the longitudinal component of the wave with the wavelength λ_2 and that, with the same precision, the transversal components of the total electromagnetic field will be proportional to

$$-\tfrac{3}{5}\, cos(\omega t - 2\pi x/\lambda_1) + \tfrac{8}{5}\, cos(\omega t - 2\pi x/\lambda_2), \tag{14}$$

where, for simplicity, it is assumed that the waves have linear polarization and propagate along the x axis and that zero time has been chosen appropriately. Let us examine expression (14) at fixed t and different x. Observing that

$$\frac{2\pi}{\lambda_2} \approx \frac{2\pi}{\lambda_1}(1 - \frac{\mu^2 c^2}{2\omega^2}) = \frac{2\pi}{\lambda_1} - \frac{\mu^2 c}{2\omega},$$

one obtains

$$-\tfrac{3}{5} cos(\omega t - 2\pi x/\lambda_1) + \tfrac{8}{5} cos(\omega t - 2\pi x/\lambda_1 + \mu^2 c x/2\omega).$$

By using standard trigonometric formulae, one can cast the above expression into the following form:

$$\xi \cdot cos(\omega t - 2\pi x/\lambda_1 + \phi_0),$$

where

$$\xi \equiv \tfrac{8}{5}\sqrt{(\,cos(\mu^2 c x/2\omega) - \tfrac{3}{8}\,)^2 + sin^2(\mu^2 c x/2\omega)}$$

and

$$cos\,\phi_0 \equiv \tfrac{8}{5}\,(\,cos(\mu^2 c x/2\omega) - \tfrac{3}{8}\,) \cdot \xi^{-1}$$

$$sin\,\phi_0 \equiv \tfrac{8}{5}\, sin(\mu^2 c x/2\omega) \cdot \xi^{-1}.$$

Since the size of the measuring equipment is sure to be much smaller than the distance to the source, ϕ_0 will be practically a constant phase shift and therefore can be neglected. One can also see that the *intensity* of the radiation considered will differ by the factor ξ^2 from the intensity of the radiation the same source should have emitted according to conventional electrodynamics. At $x = 0$ this factor is unity, as it should be. As x increases, ξ^2 monotonically grows until it reaches its maximum of $(2.2)^2 = 4.84$ at $x = 2\pi\omega/\mu^2 c \equiv x_c$. After that, ξ^2 monotonically decreases to its initial unity value, which it reaches at $x = 2x_c$. The behavior of the factor ξ^2 versus x/x_c becomes obvious if one observes that

$$\xi^2 = 2.92 - 1.92 \cdot cos(\pi x/x_c). \tag{15}$$

It is shown in Figure 3. One should notice that the distance x_c depends *quadrati-*

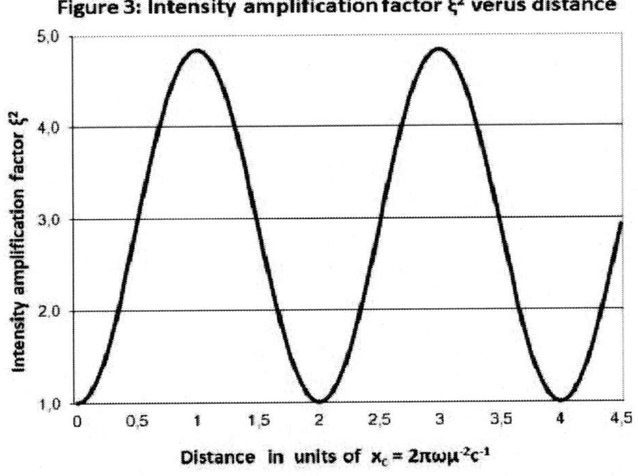

cally on the length parameter μ^{-1}. It is also inversely proportional to the radiation wavelength: $x_c = 4\pi^2/\mu^2\lambda_1$. The latter fact implies that for a given (unknown) μ, the value of x_c for radio waves is smaller than it is for infrared radiation; the value of x_c for infrared radiation is smaller than it is for visible light; and so on. In other words, the larger the wavelength, the bigger value of the length parameter μ^{-1} one is able to measure. For example, if the largest distance at which one can make measurements is of the order of 10 billion light years, then for infrared radiation with wavelength $\sim 10^{-4}$ cm, the effect can be observed if $\mu^{-1} \leq 10^{11}$ cm, whereas for radio waves with wavelength ~ 1 m, the effect will be observable if $\mu^{-1} \leq 10^{14}$ cm.

If one could record radiation coming from outer space with *any* wavelength, the considered intensity boosting effect would be observed with certainty if the bivector gauge fields associated with electromagnetism do exist and if their dynamics is indeed described by Lagrangian density (7). In reality, there are obvious limitations to one's ability to record low frequencies. To see what this means in relation to observing the effect described, let us rewrite formula (15) in the following way:

$$\xi^2 = 2.92 - 1.92 \cdot cos(\pi\omega_c/\omega), \tag{16}$$

where $\omega_c = \mu^2 cx/2\pi$ and x is the distance to the source. The behavior of ξ^2 versus ω/ω_c is shown in Figure 4. As can be seen, the effect could be observed only if the

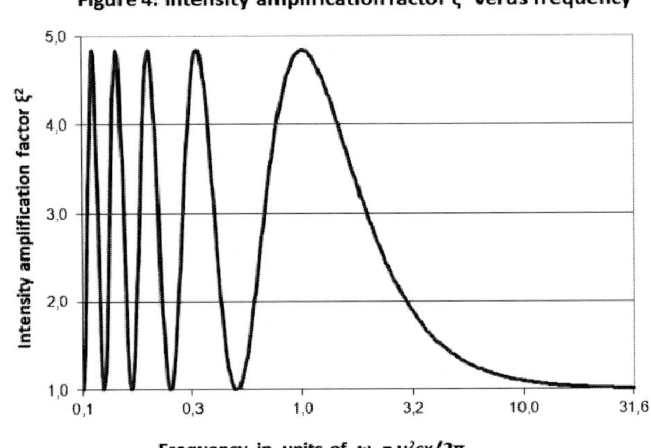

lowest frequency that can be recorded is of the order of or smaller than ω_c.

It should be emphasized that to be able to attribute the increase in intensity of radiation emitted by distant cosmic sources to the existence of bivector gauge fields, the characteristic patterns must be observed both in the variation of intensity with distance to the source for radiation with same frequency (as depicted in Figure 3) and in the variation of intensity with frequency for radiation coming from a single source (as shown in Figure 4).

References

[1] A.B.Krasulin, Five-dimensional tangent vectors in space-time. See e.g. math-ph/9805004, 9805025, 9807004, 9808006, 9808014 and 1006.3010 at arxiv.org.

[2] A.Krasulin, Bivector gauge fields from classical electrodynamics, In: *Dvoeglazov, V.V. (ed.): Photon and Poincare group*, 1999, p.218-230.

[3] L.V.Avdeev and M.V.Chizhov, *Phys. Lett. B* **321** (1994) 212.

[4] A.Krasulin, Five-vector or Bivector?, In: *Dvoeglazov, V.V. (ed.): Einstein and Others: Unification*, 2015, Nova Science Publishers, Chapter 7, pp.181-202.

Chapter 6

Inflation of the Universe by Nonlinear Electrodynamics

S. I. Kruglov *
Department of Physics, University of Toronto, Toronto, Canada
Canadian Quantum Research Center, BC, Canada

Abstract

We show that Universe inflation occurs when Einstein's gravity couples to nonlinear electromagnetic fields with cosmic stochastic magnetic fields background. Nonlinear electrodynamics with two parameters is used. The strength of magnetic fields to have the Universe inflation is obtained. It is shown that singularities of the energy density and pressure are absent for any scale factors. At large scale factors one arrives at equation of state for ultra-relativistic case. It is demonstrated that the curvature invariants do not have singularities. By computing the deceleration parameter we show that graceful exit takes place with reasonable e-folding number. The duration of Universe inflation as a function of model parameters is obtained. The classical stability and causality are analyzed by calculating the speed of the sound.

1. Introduction

The Standard Cosmological Model (SCM) is successful but it does not solve such important problems as the initial singularity and Universe acceleration. Within the SCM there are singularities of curvature invariants at the time of

*Corresponding Author's Email: kruglov@rogers.com

In: Mathematical Problems in Relativity, Gravitation, and Cosmology
Editor: Valeriy Dvoeglazov
ISBN: 979-8-89530-622-2

the creation of Universe named Big Bang. Inflation is a rapid cosmological expansion at the initial time after Big Bang. Originally, the inflationary scenario was proposed by Guth [1] to have the initial state that required by the SCM. Universe inflation can be described by different ways: deforming general relativity [2, 3, 4], by the introduction of the cosmological constant, and by using quintessence (a scalar field) [5]. Here, we describe Universe inflation with the help of nonlinear electrodynamics (NED) as the source of Einstein' gravity. Born and Infeld [6] proposed one-parameter NED which smoothes point-like charge singularity possessing self-energy finite. Also, due to quantum corrections Maxwell's electrodynamics becomes NED (Euler–Heisenberg electrodynamics) [7, 8, 9]. Therefore, it is justified to consider Einstein' gravity coupled to NED because in the inflation era electromagnetic fields were very strong. Some scenarios of Universe inflation by stochastic magnetic background were considered in [10, 11, 12, 13, 14, 15, 16, 17, 18, 19, 20, 21, 22, 23]. Here, we use NED with two parameters, which becomes Maxwell's electrodynamics at weak fields, and it drives the Universe to accelerate for the stochastic magnetic background field. Thus, our model solves problems connected with the initial singularity and Universe acceleration. In section 2 we consider Einstein's gravity coupled to NED with two parameters β and σ. We find the strength of magnetic fields when the Universe inflation occurs. It is shown that singularities of the energy density and pressure are absent. For large scale factors one has equation of state for ultra-relativistic case and graceful exit takes place with reasonable e-folding number. We demonstrate that singularities of curvature invariants are absent. We calculate the deceleration parameter in section 3 which shows the evolution of universe. The duration of the Universe inflation is analysed. We compute the speed of sound and study the causality and unitarity principles. Section 4 is a conclusion.

We use units with $c = \hbar = 1$ and metric signature is $\eta = \text{diag}(-, +, +, +)$.

2. Cosmology

The Einstein–Hilbert action coupled to NED is

$$S = \int d^4x\sqrt{-g}\left[\frac{1}{2\kappa^2}R + \mathcal{L}\right], \tag{1}$$

where R is the Ricci scalar, $\kappa^2 = 8\pi G$ and G is Newton's constant. We propose the NED Lagrangian as

$$\mathcal{L} = -\frac{\mathcal{F}}{4\pi(1 + 2\beta\mathcal{F})^{\sigma}}, \tag{2}$$

and $\mathcal{F} = F^{\mu\nu}F_{\mu\nu}/4 = (B^2 - E^2)/2$ is the Lorentz invariant, E and B are the electric and magnetic fields, correspondingly. The β is dimensional parameter and σ is dimensionless parameter. When $\beta\mathcal{F} \to 0$ Lagrangian (2) becomes Maxwell's Lagrangian. We will study Universe inflation with stochastic magnetic fields background. From action (1), we obtain equations as follow:

$$R_{\mu\nu} - \frac{1}{2}g_{\mu\nu}R = -\kappa^2 T_{\mu\nu}, \tag{3}$$

$$\nabla_\mu(\mathcal{L}_\mathcal{F}F^{\mu\nu}) = 0, \tag{4}$$

where

$$\mathcal{L}_\mathcal{F} = \frac{\partial\mathcal{L}}{\partial\mathcal{F}} = \frac{2\beta(\sigma-1)\mathcal{F}-1}{4\pi(1+2\beta\mathcal{F})^{\sigma+1}}, \tag{5}$$

and the stress-energy tensor is given by

$$T_{\mu\nu} = -F_{\mu\rho}F^{\rho}_{\ \nu}\mathcal{L}_\mathcal{F} - g_{\mu\nu}\mathcal{L}\left(\mathcal{F}\right). \tag{6}$$

We consider the line element of homogeneous and isotropic cosmological spacetime in the form

$$ds^2 = -dt^2 + a(t)^2\left(dx^2 + dy^2 + dz^2\right), \tag{7}$$

where $a(t)$ is a scale factor. The cosmic background is stochastic magnetic fields and after averaging the magnetic fields we have the isotropy of the Friedman–Robertson–Walker space-time [24]. In our approach the wavelength of electromagnetic waves is smaller than the curvature. As a result, after averaging of magnetic field we have

$$\langle\mathbf{B}\rangle = 0, \quad \langle B_i B_j\rangle = \frac{1}{3}B^2 g_{ij}. \tag{8}$$

Here, the brackets denote an average over a volume. In the following we omit the brackets. It is worth noting that the NED stress-energy tensor in such approach may be represented as for a perfect fluid [14]. For three dimensional flat universe the Friedmann's equation is given by

$$3\frac{\ddot{a}}{a} = -\frac{\kappa^2}{2}\left(\rho + 3p\right), \tag{9}$$

with $\ddot{a} = \partial^2 a/dt^2$. The Universe accelerates when $\rho + 3p < 0$. By virtue of Eq. (6) we find the energy density and pressure

$$\rho = -\mathcal{L} = \frac{B^2}{8\pi(1+\beta B^2)^\sigma},$$

$$p = \mathcal{L} - \frac{2B^2}{3}\mathcal{L}_{\mathcal{F}} = \frac{B^2[1-\beta B^2(4\sigma-1)]}{24\pi(1+\beta B^2)^{\sigma+1}}. \quad (10)$$

From Eq. (10) one obtains

$$\rho + 3p = \frac{B^2[1-\beta B^2(2\sigma-1)]]}{4\pi(1+\beta B^2)^{\sigma+1}}. \quad (11)$$

Making use of Eq. (11) and the requirement $\rho + 3p < 0$ to have the Universe acceleration, we find

$$\sqrt{\beta}B > \frac{1}{\sqrt{2\sigma-1}}. \quad (12)$$

According to Eq. (12) we have the restriction $\sigma > 1/2$. The plot of the function $\sqrt{\beta}B$ versus σ is depicted in Fig. 1. The conservation of the stress-energy tensor, $\nabla^\mu T_{\mu\nu} = 0$, gives the equation

$$\dot{\rho} + 3\frac{\dot{a}}{a}(\rho + p) = 0. \quad (13)$$

With the help of Eq. (10), we obtain

$$\rho + p = \frac{B^2[1+\beta B^2(1-\sigma)]}{6\pi(1+\beta B^2)^{\sigma+1}}. \quad (14)$$

Making use of Eqs. (13) and(14), one finds

$$B(t) = \frac{B_0}{a(t)^2}. \quad (15)$$

The B_0 is the magnetic field when $a(t) = 1$. It was shown in Ref. [14] that Eq. (15) takes place for any NED Lagrangians. Because the scale factor increases during the inflation, the magnetic field decreases. From Eqs. (10) and (15) we obtain

$$\lim_{a(t)\to\infty} \rho(t) = \lim_{a(t)\to\infty} p(t) = 0,$$

$$\lim_{a(t)\to 0} \rho(t) = \lim_{a(t)\to 0} p(t) = \infty \qquad \sigma < 1,$$

$$\lim_{a(t)\to 0} \rho(t) = \lim_{a(t)\to 0} p(t) = 0 \qquad \sigma > 1,$$

$$\lim_{a(t)\to 0} \rho(t) = -\lim_{a(t)\to 0} p(t) = \frac{1}{8\pi\beta} \qquad \sigma = 1. \quad (16)$$

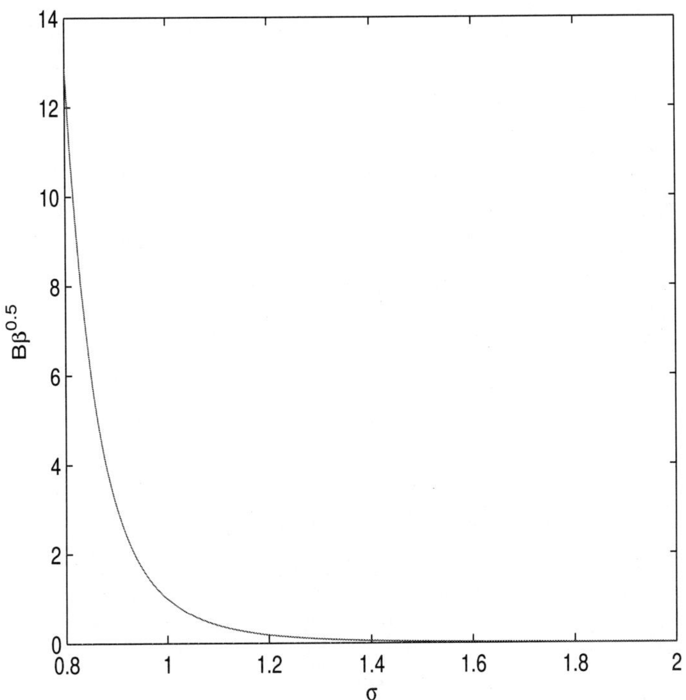

Figure 1. The function $\sqrt{\beta}B$ vs. σ. As $\sigma \to 1/2$, $B \to \infty$ and when $\sigma \to \infty$, one has $B \to 0$.

Thus, singularities of the energy density and pressure as $a(t) \to 0$ are absent at $\sigma \geq 1$. At the beginning of the Universe evolution when $a \approx 0$, we have at $\sigma = 1$ $\rho = -p$ that corresponds to de Sitter space-time. By virtue of Eq. (10) one finds the equation of state

$$w = p(t)/\rho(t) = \frac{1 - (4\sigma - 1)\beta B^2}{3(1 + \beta B^2)}. \tag{17}$$

The function w versus βB^2 is given in Fig. 2 corresponding to $\sigma = 0.75,\ 1,\ 1.5$. From Eq. (17) we have

$$\lim_{B \to 0} w = \frac{1}{3}, \tag{18}$$

$(a(t) \to \infty)$ that corresponds to the equation of state for ultra-relativistic case [25]. At $\beta B^2 = 1/(\sigma - 1)$ de Sitter spacetime occurs, $w = -1$. Making use

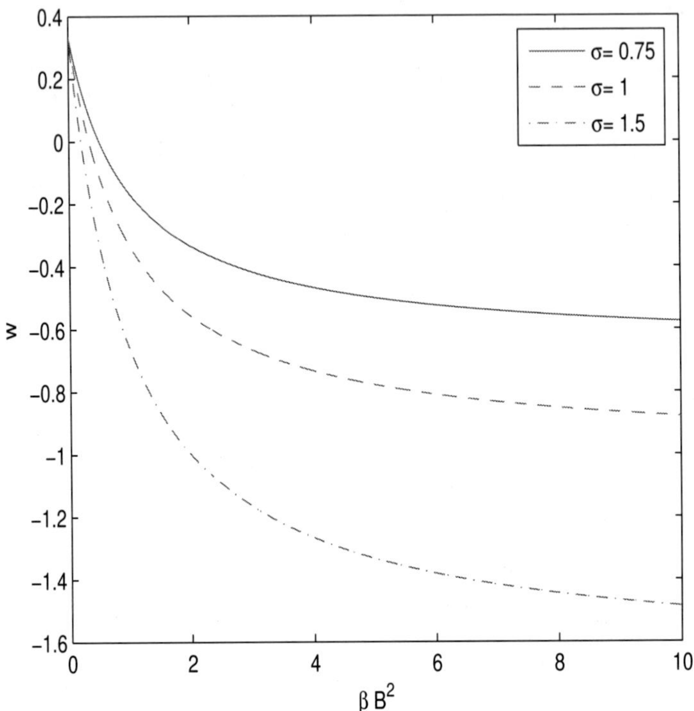

Figure 2. The function w vs. βB^2 for $\sigma = 0.75, 1, 1.5$.

of Eqs. (3) and (6), we find the Ricci scalar

$$R = \kappa^2 T_\mu^{\ \mu} = \frac{\kappa^2 \sigma \beta B^4}{2\pi(1+\beta B^2)^{\sigma+1}} = \kappa^2(\rho - 3p). \tag{19}$$

We depict the function $\beta R/\kappa^2$ versus βB^2 in Fig. 3. By virtue of Eq. (19) we obtain

$$\lim_{B\to 0} R(t) = 0,$$
$$\lim_{B\to\infty} R(t) = 0 \quad \text{at} \quad \sigma > 1,$$
$$\lim_{B\to\infty} R(t) = \frac{\kappa^2}{2\pi\beta} \quad \text{at} \quad \sigma = 1. \tag{20}$$

The Ricci tensor squared $R_{\mu\nu}R^{\mu\nu}$ and the Kretschmann scalar $R_{\mu\nu\alpha\beta}R^{\mu\nu\alpha\beta}$ are expressed in the form of combinations of $\kappa^4\rho^2$, $\kappa^4\rho p$, and $\kappa^4 p^2$ [17].

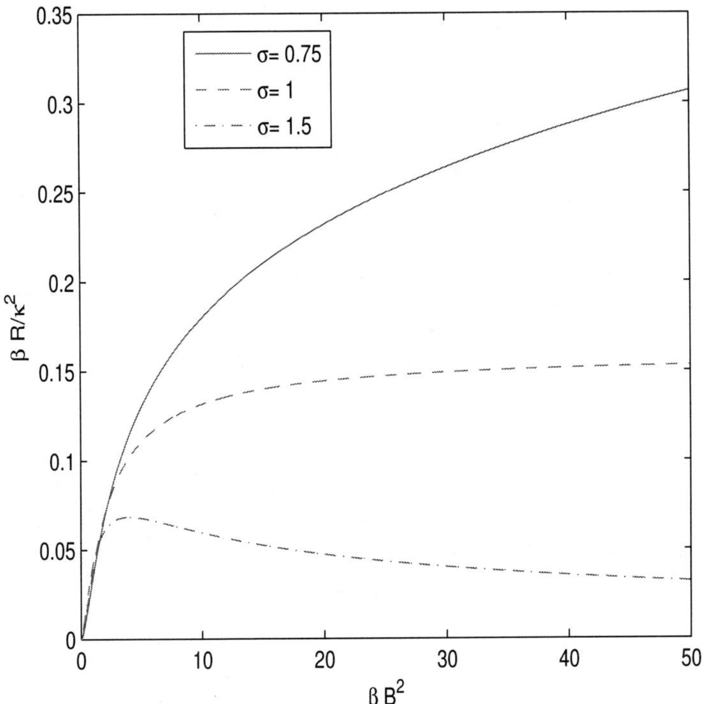

Figure 3. The function $\beta R/\kappa^2$ vs. βB^2 for $\sigma = 0.75,\ 1,\ 1.5$. When $\sigma \geq 1$ the Ricci scalar singularity as $B \to \infty$ is absent.

Therefore, in accordance with Eq. (16), they are finite as $a(t) \to 0$ and $a(t) \to \infty$ for $\sigma \geq 1$. Equations (12) and (15) show that Universe inflation takes plase at $a(t) < \sqrt[4]{\beta B_0^2(2\sigma - 1)}$.

3. Universe evolution

To study the Universe evolution we consider the Friedmann equation for three dimensional flat Universe

$$\left(\frac{\dot{a}}{a}\right)^2 = \frac{\kappa^2 \rho}{3}. \tag{21}$$

With the help of Eqs. (10) and (21), one finds

$$\dot{a} = \frac{\kappa B_0 a^{2\sigma-1}}{2\sqrt{6\pi}(a^4 + \beta B_0^2)^{\sigma/2}}. \tag{22}$$

Introducing the unitless variables $x = a/(\beta^{1/4}\sqrt{B_0})$, $y = 2\sqrt{6\pi\beta}\dot{x}/\kappa$ Eq. (22) reads

$$y = \frac{x^{2\sigma-1}}{(x^4 + 1)^{\sigma/2}}. \tag{23}$$

We depicted the function $y(x)$ in Fig. 4. Figure 4 shows that the inflation

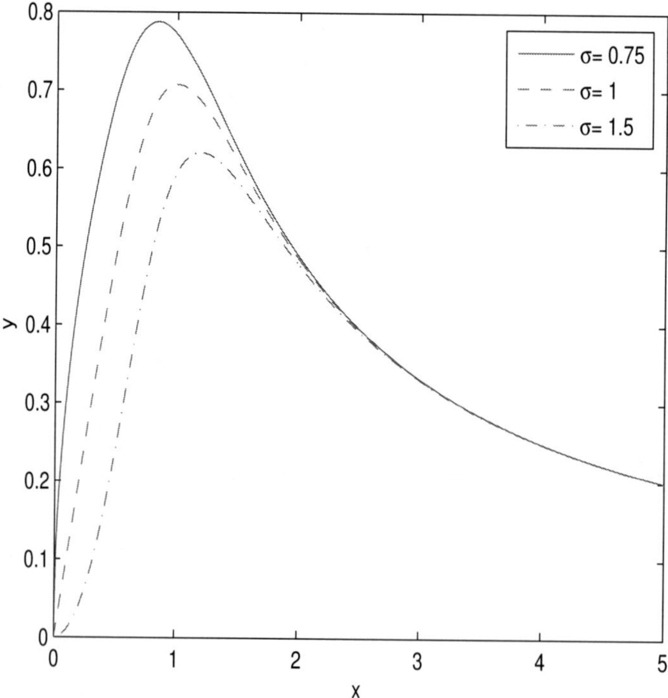

Figure 4. The function $y = 2\sqrt{6\pi\beta}\dot{x}/\kappa$ vs. $x = a/(\beta^{1/4}\sqrt{B_0})$ for $\sigma = 0.75,\ 1,\ 1.5$.

starting from Big Bang possesses the graceful exit when $\ddot{a} = 0$ $(dy/dx = 0)$, $x_{end} = \sqrt[4]{2\sigma - 1}$. Then the Universe decelerates. In accordance with Fig. 4

when σ increases the graceful exit point x_{end} (a_{end}) also increases. Making use of Eq. (22) we obtain

$$\int_{a(t_{in})}^{a(t_{end})} \frac{(a^4 + \beta B_0^2)^{\sigma/2}}{a^{2\sigma-1}} da = \frac{\kappa B_0}{2\sqrt{6}\pi} \int_{t_{in}}^{t_{end}} dt. \tag{24}$$

Integrating Eq. (24) one arrives at the equation

$$\frac{a(t_{end})^{2(1-\sigma)}(\beta B_0^2)^{\sigma/2}}{2(1-\sigma)} F\left(\frac{1-\sigma}{2}, -\frac{\sigma}{2}; \frac{3-\sigma}{2}; -\frac{a(t_{end})^4}{\beta B_0^2}\right)$$

$$-\frac{a(t_{in})^{2(1-\sigma)}(\beta B_0^2)^{\sigma/2}}{2(1-\sigma)} F\left(\frac{1-\sigma}{2}, -\frac{\sigma}{2}; \frac{3-\sigma}{2}; -\frac{a(t_{in})^4}{\beta B_0^2}\right) = \frac{\kappa B_0}{2\sqrt{6}\pi} \Delta t. \tag{25}$$

The $F(a, b; c; z)$ is the hypergeometric function and $\triangle t = t_{end} - t_{in}$ is the duration of the inflation. The hypergeometric function in Eq. (25) obeys the relation $c - a = 1$, and can be expressed in the form of the incomplete B-function, $B_z(p, q) = p^{-1} z^p F(1 - q, p; p + 1; z)$ [26]. With the help of Eq. (25) we will study the Universe inflation evolution.

Making use of Eqs. (9), (10), (15) and (21) we find the deceleration parameter which describes the Universe expansion

$$q = -\frac{\ddot{a}a}{(\dot{a})^2} = \frac{\rho + 3p}{2\rho} = \frac{x^4 + 1 - 2\sigma}{x^4 + 1}. \tag{26}$$

The deceleration parameter q versus $x = a/(\beta B_0^2)^{1/4}$ is plotted in Fig. 5. The inflation occurs when $q < 0$ and stops at the graceful exit $q = 0$ ($x_{end} = \sqrt[4]{2\sigma - 1}$). Then the deceleration phase starts ($q > 0$). It is worth noting that a singularity at the early epoch is absent. To evaluate the amount of the inflation we use the e-folding number [27]

$$N = \ln \frac{a(t_{end})}{a(t_{in})}. \tag{27}$$

Here, t_{end} is the inflation final time and t_{in} is an initial time of the inflation. The point of graceful exit is $x_{end} = \sqrt[4]{2\sigma - 1}$ and $a(t_{end}) = \sqrt[4]{2\sigma - 1}\beta^{1/4}\sqrt{B_0}$. The horizon and flatness problems may be solved if the e-folding number is $N \approx 70$ [27]. By virtue of Eq. (27) one finds the scale factor which corresponds to the inflation initial time

$$a(t_{in}) = \frac{\sqrt[4]{2\sigma - 1}\beta^{1/4}\sqrt{B_0}}{\exp(70)} \approx 4\sqrt[4]{2\sigma - 1} \times 10^{-31}\beta^{1/4}\sqrt{B_0}. \tag{28}$$

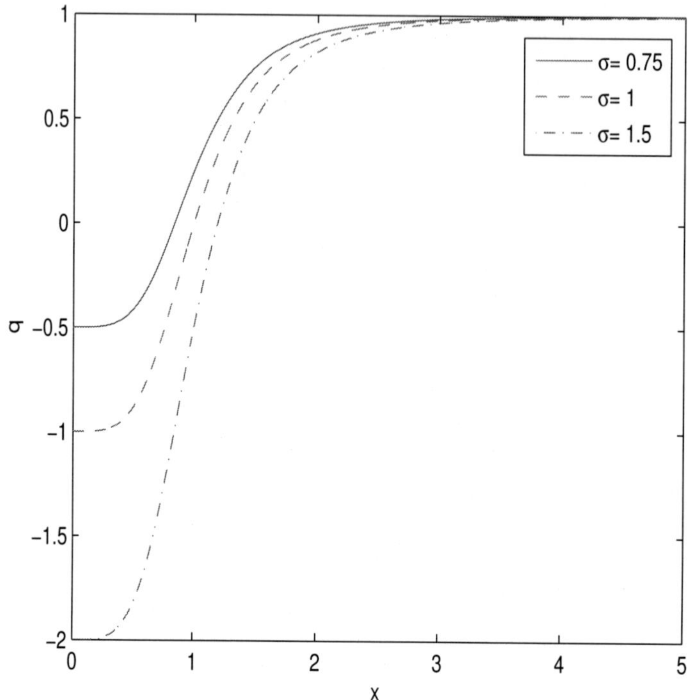

Figure 5. The function q vs. $x = a/(\beta B_0^2)^{1/4}$.

Our model has phases of the Universe inflation, the graceful exit and deceleration. From Eq. (25) and inequality $a(t_{end}) \gg a(t_{in})$ $(a(t_{end}) \approx 4 \times 10^{31} a(t_{in}))$, we obtain

$$\triangle t = \frac{\sqrt{6\pi} a(t_{end})^{2(1-\sigma)} \beta^{\sigma/2} B_0^{\sigma-1}}{(1-\sigma)\kappa} F\left(\frac{1-\sigma}{2}, -\frac{\sigma}{2}; \frac{3-\sigma}{2}; -\frac{a(t_{end})^4}{\beta B_0^2}\right). \tag{29}$$

At $a(t_{end}) \gg 1$, one can use the Pfaff transformation [26]

$$F(a, b; c; z) = (1-z)^{-a} F\left(a, c-b; c; \frac{z}{z-1}\right). \tag{30}$$

With the aid of Eq. (30), at $a(t_{end}) \gg 1$, we find

$$F\left(\frac{1-\sigma}{2}, -\frac{\sigma}{2}; \frac{3-\sigma}{2}; -\frac{a(t_{end})^4}{\beta B_0^2}\right) = F\left(-\frac{\sigma}{2}, \frac{1-\sigma}{2}; \frac{3-\sigma}{2}; -\frac{a(t_{end})^4}{\beta B_0^2}\right)$$

$$\approx \left(1+\frac{a(t_{end})^4}{\beta B_0^2}\right)^{\sigma/2} F\left(-\frac{\sigma}{2},1;\frac{3-\sigma}{2};1\right) = (1-\sigma)\left(1+\frac{a(t_{end})^4}{\beta B_0^2}\right)^{\sigma/2}. \tag{31}$$

We have used formulas [28] $F(a,b;c;1) = \Gamma(c)\Gamma(c-a-b)/(\Gamma(c-a)\Gamma(c-b))$, $\Gamma(1+z) = z\Gamma(z)$. Making use of Eqs. (29) and (31), at $a(t_{end}) \gg 1$, one finds the equation

$$a(t_{end}) = \sqrt{\frac{\kappa B_0 \Delta t}{\sqrt{6\pi}}}, \tag{32}$$

corresponding to the radiation era. From Eq. (32), $a(t_{end}) = \sqrt[4]{(2\sigma-1)\beta B_0^2}$, the coupling β is given by

$$\beta = \frac{\kappa^2 \Delta t^2}{6\pi(2\sigma-1)}. \tag{33}$$

If one takes the inflation duration $\triangle t \approx 10^{-32}s$, the values of $\kappa = \sqrt{8\pi G}$, one can find β from Eq. (33). The function $y = \beta/(\kappa^2\Delta t^2)$ versus σ is depicted in Fig. 6. Figure 6 shows that when parameter σ increases the coupling β decreases in order to have the same Universe inflation time.

3.1. Causality and Unitarity

When the sound speed is less than the light local speed ($c_s \leq 1$) [29] the causality takes place, and when the sound square speed is positive ($c_s^2 > 0$) a classical stability holds. By virtue of Eq. (10) we obtain the speed squared of sound

$$c_s^2 = \frac{dp}{d\rho} = \frac{(\beta B^2)^2(4\sigma^2-5\sigma+1)+\beta B^2(2-9\sigma)+1}{3(1+\beta B^2)(1+\beta B^2(1-\sigma))}. \tag{34}$$

The function c_s^2 versus βB^2 is plotted in Fig. 7. In accordance with Fig. 7 when parameter σ increases the causality and unitarity hold for smaller background magnetic field. From Eq. (34) we obtain that $c_s^2 = 0$ at

$$\beta B^2 = \frac{9\sigma-2\pm\sqrt{\sigma(65\sigma-16)}}{2(4\sigma^2-5\sigma+1)}. \tag{35}$$

By virtue of Eqs. (12) and (35) and Fig. 7, we find that at the acceleration phase, the classical stability, causality and unitarity are broken.

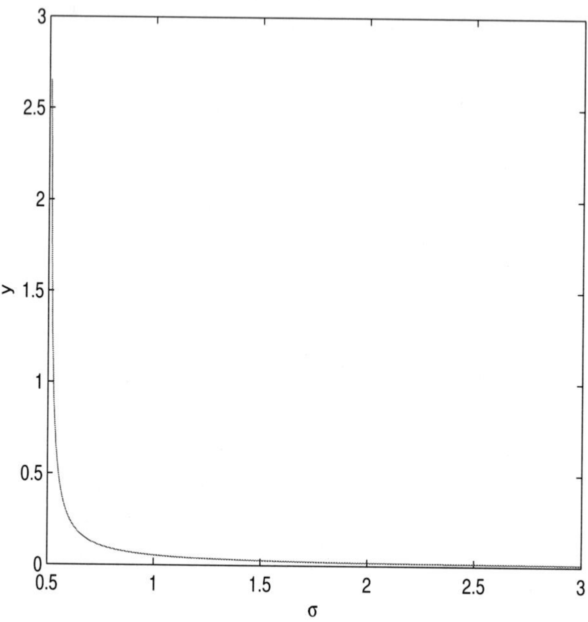

Figure 6. The function $y = \beta/(\kappa^2 \Delta t^2)$ vs. σ.

Conclusion

Einstein's gravity coupled to NED with two parameters has been studied. We showed that the background stochastic magnetic fields, as a source of gravity, leads to Universe inflation. The magnetic fields interval when the Universe inflation occurs has been found. The singularities of the energy density and pressure are absent in our model. It was shown that at the large scale factor the equation of state corresponds to ultra-relativistic case. We have demonstrated that there are not singularities of curvature invariants. The evolution of the universe has ben described. We have obtained the function of the scale factor on the time. The deceleration parameter has been calculated to study the Universe evolution. We have analysed the amount of the inflation by considering the e-folding number. It was demonstrated that for some model parameters the reasonable e-folding number takes place. The Universe inflation duration as the function of parameters β and σ has been calculated. To study the causality and unitarity principles we computed the speed of sound. We have obtained the interval of background magnetic fields when causality and unitarity hold. There are similarities and differences of our Universe inflation

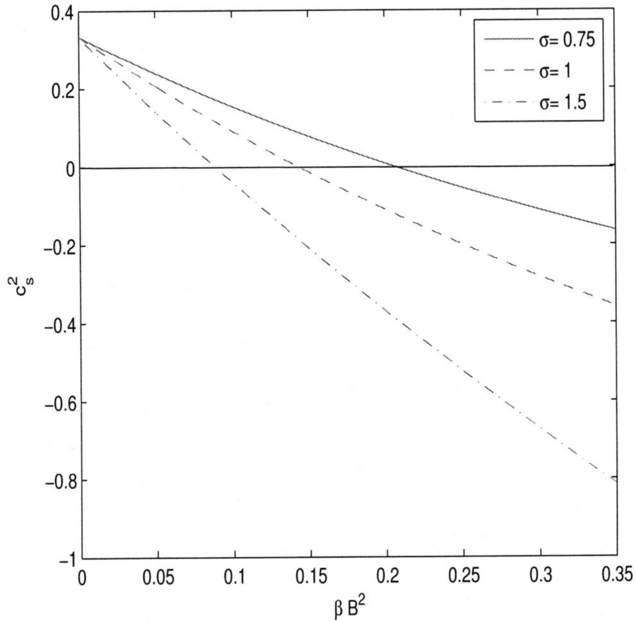

Figure 7. The function c_s^2 vs. βB^2.

model compared with models in Refs. [30, 31]. The SCM problems of the initial singularity and Universe acceleration are solved in our model.

References

[1] A. H. Guth, *Phys. Rev. D* **23**, 347 (1981).

[2] Alexei A. Starobinsky, *Phys. Lett. B* **91** 99-102 (1980).

[3] S. Capozziello and V. Faraoni, *Beyond Einstein Gravity: A Survey of Gravitational Theories for Cosmology and Astrophysics* (Springer Science+Business Media B.V., New York, 2011).

[4] S. Nojiri and S. D. Odintsov, *Phys. Rept.* **505**, 59 (2011).

[5] Andrei D. Linde, *Lect. Notes Phys.* **738**, 1-54 (2008).

[6] M. Born and L. Infeld, *Proc. Royal Soc.* (London) A **144**, 425 (1934).

[7] W. Heisenberg and E. Euler, *Z. Physik* **98**, 714 (1936).

[8] J. Schwinger, *Phys. Rev.* **82**, 664 (1951).

[9] S. L. Adler, *Ann. Phys.* (N.Y.) **67**, 599 (1971).

[10] C. S. Camara, M. R. de Garcia Maia, J. C. Carvalho, and J. A. S. Lima, *Phys. Rev. D* **69**, 123504 (2004).

[11] E. Elizalde, J. E. Lidsey, S. Nojiri, and S. D. Odintsov, *Phys. Lett. B* **574**, 1 (2003).

[12] V. A. De Lorenci, R. Klippert, M. Novello, and J. S. Salim, *Phys. Rev. D* **65**, 063501 (2002).

[13] M. Novello, S. E. Perez Bergliaffa, and J. M. Salim, *Phys. Rev. D* **69**, 127301 (2004).

[14] M. Novello, E. Goulart, J. M. Salim, and S. E. Perez Bergliaffa, *Class. Quant. Grav.* **24**, 3021 (2007).

[15] D. N. Vollick, *Phys. Rev. D* **78**, 063524 (2008).

[16] R. García-Salcedo, T. Gonzalez, A. Horta-Rangel, and I. Quiros. *Phys. Rev. D* **90**, 128301 (2014).

[17] S. I. Kruglov, *Phys. Rev. D* **92**, 123523 (2015).

[18] S. I. Kruglov, *Int. J. Mod. Phys. A* **35**, 2050168 (2020).

[19] S. I. Kruglov, *Eur. Phys. J. Plus* **135**, 370 (2020).

[20] S. I. Kruglov, *Int. J. Mod. Phys. D* **29**, 2050102 (2020).

[21] S. I. Kruglov, *Int. J. Mod. Phys. A* **32**, 1750071 (2017).

[22] S. I. Kruglov, *Int. J. Mod. Phys. A* **31**, 1650058 (2016).

[23] S. I. Kruglov, *Int. J. Mod. Phys. D* **25**, 1640002 (2016).

[24] R. Tolman and P. Ehrenfest, *Phys. Rev.* **36**, 1791 (1930).

[25] L. D. Landau and E. M. Lifshits, *The Classical Theory of Fields* (Pergamon Press, 1975).

[26] H. Bateman and A. Erdelyi, *Higher Transcendental Functions* (McGraw-Hill Book Company, Inc, 1953).

[27] A. R. Liddle and. H. Lyth, *Cosmological Inflation and Large-Scale Structure* (Cambrige University Press, 2000).

[28] Handbook of Mathematical Functions with Formulas, Graphs and Mathematical Tables, Edit by M. Abramowitz and I. Stegun. National Bureau of Standarts, *Applied Mathematics Series 55* (1972).

[29] R. García-Salcedo, T. Gonzalez, and I. Quiros, *Phys. Rev. D* **89**, 084047 (2014).

[30] H. B. Benaoum1, Genly Leon, A. Övgün, and H. Quevedo. *Eur. Phys. J. C* **83**, 367 (2023).

[31] S. I. Kruglov, *Eur. Phys. J. C* **84**, 205 (2024).

Chapter 7

On the Non-Classical Representation of the Basic Concepts of Subatomic Physics

V. V. Varlamov *
Siberian State Industrial University, Novokuznetsk, Russia

Abstract

The possibility of constructing a quantum theory without involving classical models is discussed. The construction is based on a holistic approach to the description of quantum microphenomena, which are understood as states of a single quantum system. From the holistic viewpoint, substance (the same as energy) is the primary concept, and particles are secondary and emergent. An axiomatic system is proposed, where the basic notion of the spectrum of matter is defined. Non-classical (algebraic) analogues of the main characteristics of the states of a single quantum system, such as mass, charge, spin, and discrete symmetries, are given.

Keywords: non-locality, mass, spin, charge, discrete symmetries, holism, single quantum system, entanglement

PACS 02.20-a, 03.65.-w, 03.65.Fd.

I think we may ultimately reach the stage when it is possible to set up quantum theory without any reference to classical theory.
P.A.M. Dirac [1]

---

*Corresponding Author's Email: varlamov@sibsiu.ru

In: Mathematical Problems in Relativity, Gravitation, and Cosmology
Editor: Valeriy Dvoeglazov
ISBN: 979-8-89530-622-2

1. Introduction

The quotation from Dirac in the epigraph suggests that the application of classical models in quantum theory is not necessary. Dirac writes: "There is no well-defined unique process for passing from classical theory to quantum theory. That means that when we set up a quantum theory, we have to set it up to stand on its own feet, independent of the classical theory. The only value of the classical theory is to provide us with hints for getting a quantum theory; the quantum theory is then something that has to stand in its own rights. If we were sufficiently clever to be able to think of a good quantum theory straight away, we could manage without classical theory at all ... I think we may ultimately reach the stage when it is possible to set up quantum theory without any reference to classical theory" [1, P. 42-43]. The construction of a quantum theory should be carried out exclusively using its mathematical apparatus. The main premise of the algebraic formulation of quantum theory is the possibility of constructing a theory without involving any classical analogies or visual images and mechanical models associated with these analogies. Any macroscopic analogies introduced from classical physics should be discarded. As for subatomic physics, the construction of the theory should mainly be based on a group-theoretic (symmetric) approach.

On the other hand, consideration of quantum microphenomena is impossible without accepting their wholeness, and the wholeness is initially undivided into parts [2]. The ideas that everything is built of particles, the whole is the sum of its parts, and the global properties of a system S are entirely determined by the states and interactions of its parts, make up the core of *reductionism* and the *separability principle*. In a reductionist model, a system is necessarily treated as a mechanical sum of its parts. Although such a mechanical approach is invalid at the quantum level, it is still applied both at the atomic (the Bohr planetary model) and even subatomic (the quark model) scales.

The fact that separability (which is the main principle of reductionism) is only partly applicable in quantum mechanics is well known. If some parts of a system S are entangled (a nonseparable state of S), then none of the global properties of S are determined by the properties of its parts. On the other hand, no part of S can be in a pure state, i.e., none of the subsystems S_1, S_2, . . ., S_N of S can exist independently. In no case can the parts of an entangled quantum system be treated as autonomous. In other words, the parts of a system in a nonseparable state dissolve themselves in the whole, thus depriving it of a structure in the ordinary sense. This does not mean that the system is completely amorphous, but only that its structure is not borrowed from the alien âĂIJrepertoire of classical physicsâĂİ (as Heisenberg called it) but naturally stems from the mathematics of quantum theory, such as quantum state vectors, symmetry groups, Hilbert space, tensor products of Hilbert and $\mathbb{K}$-Hilbert spaces, etc. This study is an attempt to find such a structure.

In recent times, reductionist ideology has continuously given ground to its antithesis, the holistic approach. It becomes evident that quantum phenomena are essentially holistic, and quantum systems are naturally nonseparable. This key fact is confirmed by numerous experimental tests of Bell's inequalities reported by Friedman-Clauser, Aspect, Greenberg-Horn-Zeilinger, and others. The 2022 Nobel Prize in Physics was awarded to Alain Aspect, John F. Clauser, and Anton Zeilinger "for experiments with entangled photons, establishing the violation of Bell inequalities and pioneering quantum information science". Quantum mechanics is nonlocal and, hence, needs no reference to space-time. Since particle physics operates entirely in the field of quantum phenomena, one does not need space-time to construct a particle model and should replace the traditional localization in space-time with a purely holistic framework.

This article considers a non-classical (algebraic) description of the main characteristics of elementary particles, such as mass, spin, charge, and discrete symmetries. Elementary particles are presented within the framework of a holistic vision, i.e., as states of a single quantum system obtained as a result of the application of the Gelfand-Naimark-Segal construction. Paragraph 2 provides the basic axiomatics within the framework of the two-level Heisenberg-Fock concept, which, in a sense, separates the micro and macro worlds. The definition of the physical $\mathbb{K}$-Hilbert space and its structure are given. In paragraph 3, algebraic analogues of the main characteristics of "elementary particles" (states of a single quantum system) are considered: mass as a tensor structure of cyclic vectors of $\mathbb{K}$-Hilbert space, charge as a $\mathbb{K}$-linear structure, spin as doubling, and discrete symmetries as automorphisms of a $\mathbb{K}$-linear structure (Clifford algebras). Paragraph 4 provides an algebraic description of the effect of entanglement of states of a single quantum system.

2. Axiomatics

According to the Heisenberg-Fock concept, reality has a two-level structure: *potential reality* and *actual reality*. Heisenberg argued that any quantum microobject belongs to both sides of reality: first, potential reality as a superposition, and, secondly, actual reality after the reduction of superposition, i.e., measurement. Measurement is the localization (actualization) of any state of a single quantum system $\mathbf{U}$[1].

---

[1]Wheeler noted that no elementary phenomenon is a phenomenon until it is an observable (recorded) phenomenon. The idea that a quantum microobject exists by itself before any measurement in the form of a certain corpuscle (point-shaped or occupying a certain region $\mathcal{O}$ of space-time), the so-called "asymptotically stable localization center", is petitio principii. Moreover, the concept of the multiplicity of quantum microobjects is also a logical error. There is no multiplicity of quantum microobjects; there is a single quantum system $\mathbf{U}$, the states of which are identified as quantum microobjects (elementary particles) as a result of measurement. In fact, here lies (within the framework of $\mathbf{U}$) the solution to the problem of multiplicity and

Following Heisenberg, we assume that the major observable at the fundamental microscopic level is energy, which is quantized in accordance with Planck's law. Note that the continuum concept is out of play from the very beginning. The fundamental level is understood as one characterized by the elementary length $l_0 \sim 10^{-13}$ cm scale[2]. A particle is interpreted as a quantum of energy and described by an algebraic quantization procedure known as the Gelfand-Naimark-Segal (GNS) construction, where the energy operator H is considered as a C^*-algebra operator. We use the following system of definitions (axioms)[3] [4]:

A.I (Energy and fundamental symmetry) *A single quantum system* U *at the fundamental level is characterized by a* C^**-algebra* $\mathfrak{A}$ *with a unit consisting of the energy operator* H *and the generators of the fundamental symmetry group* G_f *attached to* H*, forming a common system of eigenfunctions with* H.

A.II (States) *A physical state of* C^**-algebra* $\mathfrak{A}$ *is determined by the cyclic vector* $|\Phi\rangle$ *of the* C^**-algebra representation* π *on separable Hilbert space* H_∞*:*

$$\omega_\Phi(H) = \frac{\langle\Phi \mid \pi(H)\Phi\rangle}{\langle\Phi \mid \Phi\rangle}.$$

The set $PS(\mathfrak{A})$ *of all pure states of* C^**-algebra* $\mathfrak{A}$ *coincides with the set of all states* $\omega_\Phi(H)$ *associated with all irreducible cyclic representations* π *of* $\mathfrak{A}$, $|\Phi\rangle \in \mathsf{H}_\infty$ *(the Gelfand–Naimark–Segal construction).*

A.III (Space of rays) *The set of all positive pure states* $\omega_\Phi(H) \geq 0$ *forms a physical* $\mathbb{K}$*-Hilbert space* $\mathbf{H}_{\rm phys}(\mathbb{K})$ *(a GNS-Hilbert space equipped with the* $\mathbb{K}$*-structure of* $*$*-ring). For each state vector* $|\Psi\rangle \in \mathbf{H}_{\rm phys}$*, there is a unit ray* $\boldsymbol{\Psi} = e^{i\alpha}|\Psi\rangle$*, where* α *runs through all real numbers and* $\sqrt{\langle\Psi\,|\Psi\rangle} = 1$.

---

identity.

[2]This level is fundamental not because of its minimality. As early as 1968, A.D. Sakharov wrote in the article "Is there an elementary length?" [3]: "So, a combination of theoretical and experimental arguments forces us to admit that the boundary of the $l_0 = r$ theory proposed by Heisenberg should be pushed towards much higher energies". Since then, this boundary has been pushed back into an even deeper range of energies $\ll l_0$. The main goal of the Large Hadron Collider (LHC) was the search for new physics (i.e., new particles) at energy levels $\ll l_0$. However, all the numerous experiments have not revealed any new physics at these levels $\ll l_0$: no supersymmetry, no dark matter particles, nothing at all that could point beyond the standard model. The l_0 level is fundamental due to the fact (now irrefutably established experimentally) that everything that exists (mainly stable matter) manifests from the primary substrate (protomatter) at energy levels $\sim l_0$.

[3]This system in no way claims to be exhaustive and complete, being in fact the first outline of this kind. Axioms **A.I–A.IV** describe the fundamental level (quantum domain), axioms **B.I** and **B.II** belong to the classical (local) domain (actual reality). The axiom of reduction **B.0** serves as a kind of "boundary between two worlds". It should be noted that all definitions, one way or another related to the mode of potency, have a pronounced apophatic character: a *non-local* quantum substrate, *non-separable* Hilbert space, *non-Boolean* logic, etc.

The ray space is the quotient space $\hat{H} = \mathbf{H}_{\text{phys}}/S^1$*, that is, the projective space of one-dimensional subspaces of* $\mathbf{H}_{\text{phys}}$*. All states of a single quantum system* $\mathbf{U}$ *are described by the unit rays.*

A.IV (Axiom of spectrality) *Space* $\hat{H}$ *contains a complete set of states with non-negative energy.*

A.V (Superposition principle) *The basic correspondence between physical states and elements of space* $\hat{H}$ *involves the superposition principle of quantum theory; that is, there is a set of basic states such that arbitrary states can be constructed from them using linear superpositions.*

B.0 (The principle of reduction) *The reduction of the superposition of rays (measurement) of the system* $\mathbf{U}$ *is described as a transition from* $\mathbf{\Psi} = \sum_k c_k \mathbf{\Psi}_k$ *to* $\mathbf{\Psi}_1$*, i.e.*

$$\mathbf{\Psi} = \sum_k c_k \mathbf{\Psi}_k \longrightarrow \mathbf{\Psi}_1$$

with probability $|c_1|^2$ *(in accordance with Born's rules).*

B.I (Localization) *Each bounded open set* $\mathcal{O}$ *of Minkowski spacetime* $\mathcal{M} = \mathbb{R}^{1,3}$ *is associated with a* C^**-algebra* $\mathfrak{A}(\mathcal{O})$ *with a unit, which is a subalgebra of a nonlocal algebra* $\mathfrak{A}$ *and is called the algebra of local observables associated with the set* $\mathcal{O}$*; in this case, the algebra* $\mathfrak{A}_{\text{loc}}$ *is a complement according to the norm of the union*

$$\mathfrak{A}_{\text{loc}} = \bigcup_{\mathcal{O}\subset\mathcal{M}} \mathfrak{A}(\mathcal{O}),$$

called the algebra of local observables. In addition, the following conditions are met:
(a) isotony: if $\mathcal{O}_1 \subset \mathcal{O}_2$*, then* $\mathfrak{A}(\mathcal{O}_1) \subset \mathfrak{A}(\mathcal{O}_2)$*;*
(b) local commutativity: algebras $\mathfrak{A}(\mathcal{O}_1)$ *and* $\mathfrak{A}(\mathcal{O}_2)$*, associated with spatially separated regions* $\mathcal{O}_1$ *and* $\mathcal{O}_2$*, commute with each other.*

B.II (Poincaré invariance) *In the Hilbert space* $\mathcal{H}$*, a unitary representation* $(a, \Lambda) \to U(a, \Lambda)$ *of the spinor Poincaré group* $\mathfrak{P}_0$ *(continuous in weak operator topology) is defined, which determines the transformation law for translations and Lorentz transformations for local observables*

$$A \longrightarrow U(a, \Lambda) A U(a, \Lambda)^{-1}$$

and state vectors

$$|\Phi\rangle \longrightarrow U(a, \Lambda)\, |\Phi\rangle$$

(here A is an arbitrary element of the algebra of local observables $\mathfrak{A}(\mathcal{O})$ or von Neumann algebra $\overline{\mathfrak{A}(\mathcal{O})}$, $|\Phi\rangle$ is an arbitrary vector from $\mathcal{H}$). In this case, for algebra $\mathfrak{A}(\mathcal{O})$, the covariance condition with respect to the proper Poincaré group is satisfied:

$$\alpha_{a,\Lambda}(\mathfrak{A}(\mathcal{O})) = \mathfrak{A}(\Lambda\mathcal{O} + a).$$

The generators of the representation $(a, \Lambda) \rightarrow U(a, \Lambda)$ – the operators of the total 4-momentum p^μ and the four-dimensional angular momentum $M^{\lambda\mu}$ – are self-adjoint operators in $\mathcal{H}$ attached to the von Neumann algebra $\overline{\mathfrak{A}(\mathcal{O})}$ of local observables. The spectrum of the energy-momentum operator p belongs to the closed upper light cone $\overline{V}^+$.

A-I. The first axiom defines the basic ingredients of the formalism. According to von Neumann [5], the primary (undefinable) concepts of quantum mechanics are system, observable, and state. As an observable, we take a C^*-algebra consisting of the energy operator H and the adjoint generators of the fundamental symmetry group G_f. This group G_f provides the structure of energy levels since the spaces of irreducible representations of G_f coincide with the eigenspaces of the energy operator H. In other words, group G_f specifies a template for the spectrum-generating mechanism provided by the GNS construction. For C^*-algebra of observables $\mathfrak{A}$ and a set of states $S(\mathfrak{A})$, we identify group G_f as the fundamental symmetry group of $\mathfrak{A}$ if the map $g \rightarrow (\alpha_g, \alpha'_g)$ of group G_f into the group of all symmetries of $(\mathfrak{A}, S(\mathfrak{A}))$ is a homomorphism, where bijections $\alpha : \mathfrak{A} \rightarrow \mathfrak{A}$ and $\alpha' : S(\mathfrak{A}) \rightarrow S(\mathfrak{A})$ satisfy the matching condition $(\alpha'\omega)(\alpha A) = \omega(A)$ for all $A \in \mathfrak{A}, \omega \in S(\mathfrak{A})$. From here on, we assume that G_f is a non-compact Lie group (for example, a Lorentz group or a conformal group). Then, for any physical state $\omega \in S(\mathfrak{A})$ and any fixed $A \in \mathfrak{A}$, the map $g \rightarrow \omega(\alpha_g(A))$ is continuous in g. Moreover, the group G_f is unitarily or antiunitarily realized if there exists a continuous representation $g \rightarrow U_g$ of G_f by unitary or antiunitary operators (according to whether α_g are algebraic automorphisms or antiautomorphisms) in the Hilbert space H_∞ such that $\alpha_g(A) = U_g A^{(*)} U_g^{-1}$ for any $A \in \mathfrak{A}$, $g \in G$, and $A^{(*)}$ standing for A or A^* in case U_g is unitary or antiunitary, respectively. In terms of the universal covering of the fundamental group $\widetilde{G}_f = \mathbf{Pin}(p, q)$, the CPT-group is given by automorphisms and antiautomorphisms of the Clifford algebra $C\!\ell_{p,q}$ [6, 7, 8, 9, 10]. The structure of automorphism $\mathcal{A} \rightarrow \overline{\mathcal{A}}$ specifies the charge conjugation C and determines the charge distribution of physical states in the $\mathbb{K}$-Hilbert space.

A-II. The second axiom establishes the GNS mechanism of spectrum generation. The arising Hilbert space is an emergent construction whose explicit form depends on the choice of group G_f (the so-called *dressing* of the operator algebra). For any state ω on a C^*-algebra $\mathfrak{A}$, the GNS construction defines a cyclic representation π_ω of $\mathfrak{A}$ on the Hilbert space H_∞ with a cyclic vector $|\Phi\rangle$ so that $\omega(\mathfrak{a}) = \langle\Phi \mid \pi_\omega(\mathfrak{a}) \mid \Phi\rangle$, $\forall\mathfrak{a} \in \mathfrak{A}$. These conditions define a representation π_ω that is unique up to unitary equivalence (which relates cyclic vectors of different representations). Further, each state ω defines a certain representation of algebra $\mathfrak{A}$, and the resulting representation π_ω is irreducible exactly when ω is *pure*.

A-III. The third axiom defines the physical $\mathbb{K}$-Hilbert space ($\mathbb{K} = \mathbb{R}, \mathbb{C}, \mathbb{H}$), where the canonical correspondence $\omega \leftrightarrow \pi_\omega$ allows one to realize the Dyson âĂIJthree-fold wayâĂİ [11, 12], which is the symmetry between quaternion, complex, and real representations of group G_f. With the respective representations, we identify the charged ($\mathbb{C}$), neutral ($\mathbb{H}$), and purely neutral ($\mathbb{R}$) states of the spectrum of matter, while the Dyson symmetry provides the dynamic relations between spin, charge, and mass in terms of the tensor product [13].

Thus, the axioms **A.I-A.III** define a single quantum system $\mathbf{U}$ consisting of a C^*-algebra (built on the energy operator H and the group G_f generators) and a $\mathbb{K}$-Hilbert space $\mathbf{H}_{\rm phys}(\mathbb{K})$ generated by the GNS construction. The cyclic vectors of space $\mathbf{H}_{\rm phys}(\mathbb{K})$ represent all possible states of $\mathbf{U}$, thus giving the spectrum of matter. For a specific choice of group G_f, eigenvectors related to the discrete eigenvalues of H define the stationary states of $\mathbf{U}$. In particular, for $G_f = \mathrm{SO}_0(1,3)$ (Lorentz group), the spectrum of matter is the spectrum of *elementary particles* [14, 15], whose complete list is given by the Particle Data Group [16]. For $G_f = \mathrm{SO}_0(2,4)$, where $\mathrm{SO}_0(2,4)$ is a conformal group, the spectrum of matter is structured as the periodic system of *chemical elements* [17, 18, 19, 20, 21]. In this case, the stationary states of $\mathbf{U}_A$ correspond to the atoms of chemical elements. Restricting the group $\mathrm{SO}_0(2,4)$ to the subgroup $\mathrm{SO}_0(1,3)$ we get the reduction $\mathbf{U}_A \rightarrow \mathbf{U}_E$, where the resulting subsystem $\mathbf{U}_E$ again corresponds to elementary particles, this time represented by the cyclic vectors of space $\mathbf{H}_{\rm phys}(\mathbb{K})$. This fact supports WeisskopfâĂŹs conjecture that nuclear and particle physics are not separate disciplines but a single science. The quantum system $\mathbf{U}_A$ corresponds to a higher symmetry group G_f than its subsystem $\mathbf{U}_E$ and, hence, describes a higher level of the organization of matter.

2.1. Symmetries of the Spectrum of Matter

Note that the symmetries of the spectrum of matter can be classified into three types: (1) *fundamental symmetries* G_f responsible for the pure states (rays) of the quantum system $\mathbf{U}$ and the corresponding coherent subspaces of $\mathbf{H}_{\rm phys}(\mathbb{K})$; (2) *dynamic symmetries* G_d describing transitions between the states of different coherent subspaces; and (3) *gauge symmetries* G_g relating pure states within a coherent subspace. All the dynamical symmetries G_d (the so-called *internal* symmetries of compact groups $\mathrm{SU}(n)$) can be lifted to the $\mathbb{K}$-Hilbert space by means of the central extension technique [22, 23]. The result includes the $\mathrm{SU}(3)$-, $\mathrm{SU}(4)$-, and $\mathrm{SU}(5)$-systematics of hadronic spectra as special cases. The dynamic symmetries $G_d = \mathrm{SU}(n)$ relate different cyclic vectors of the $\mathbb{K}$-Hilbert space, i.e., set quantum transitions between states (levels of the spectrum of matter). It is natural to assume that the operators of group G_d and its subgroups relate states with similar characteristics; for this reason, the approximate group G_d symmetries should be called *external* rather than *internal* symmetries. In the case of the atomic system $\mathbf{U}_A$, the similar relation is provided by operators Γ_+ and Γ_- connecting the homological series of the periodic system of elements (see [19, 21]). Obviously, such an *external* description of the G_d groups turns the (*internal*) quark composition of hadrons into a fiction.

2.2. The Hilbert Space Structure

Below we show that a general vector of a physical $\mathbb{K}$-Hilbert space $\mathbf{H}_{\mathrm{phys}}(\mathbb{K})$ can be represented as a product of elementary vectors of two types. An odd/even product yields a fermionic/bosonic state, correspondingly.

Theorem 1 ([24]) *A physical $\mathbb{K}$-Hilbert space $\mathbf{H}_{\mathrm{phys}}(\mathbb{K})$ can be expanded into a direct sum of (non-zero) coherent subspaces*

$$\mathbf{H}_{\mathrm{phys}}(\mathbb{K}) = \bigoplus_{\nu\in N} \mathbf{H}^{\nu}_{\mathrm{phys}}(\mathbb{K}).$$

In this case, the superposition principle takes place in a restricted form, i.e., for coherent subspaces $\mathbf{H}^{\nu}_{\mathrm{phys}}(\mathbb{K})$. A non-zero linear combination of cyclic vectors of pure separable states is a cyclic vector of a pure separable state, provided that all the original vectors lie in the same coherent subspace $\mathbf{H}^{\nu}_{\mathrm{phys}}(\mathbb{K})$. A superposition of cyclic vectors of pure separable states from different coherent subspaces is a cyclic vector of a mixed state.

In the state-of-the-art high-energy physics, the superselection rules can be completely described by electric Q $(= Q_1)$, baryon B $(= Q_2)$, and lepton L $(= Q_3)$ charges. The electric charge is taken into account by the $\mathbb{K}$-linear structure of the space $\mathbf{H}_{\mathrm{phys}}(\mathbb{K})$, see [24]. In order to describe the entire spectrum of observed states (spectrum of matter), we introduce a 2-parameter gauge group $G = U(1)^2 \equiv U(1) \times U(1)$ with respect to the baryon B and lepton L charges. Then the total space $\mathbf{H}_{\mathrm{phys}}(\mathbb{K})$ is a sum of coherent subspaces

$$\mathbf{H}_{\mathrm{phys}}(\mathbb{K}) = \bigoplus_{b,\ell\in\mathbb{Z}} \mathbf{H}^{(b,\ell)}_{\mathrm{phys}}(\mathbb{K}), \quad \mathbb{K} = \mathbb{R}, \mathbb{C}, \mathbb{H}, \tag{1}$$

with a certain baryon b and lepton ℓ numbers. Consequently, the entire set of cyclic vectors of $\mathbf{H}_{\rm phys}(\mathbb{K})$ can be divided into subspaces of vectors of the form

$$|\mathbb{K}, b, \ell, s\rangle = \left|\mathbb{K}, b, \ell, |l - \dot{l}|\right\rangle \tag{2}$$

with given values of charge, spin, mass, baryon and lepton numbers. Each vector (2) is associated with a CPT group, see [7] and subsection 3.4.

Theorem 2 ([24]) *All cyclic vectors $|\psi\rangle$ of the physical $\mathbb{K}$-Hilbert space $\mathbf{H}_{\rm phys}(\mathbb{K})$, which define fermionic and bosonic states of $\mathbb{R}$-, $\mathbb{C}$-, and $\mathbb{H}$-subspaces, are given by a composition of the fusion and doubling operations acting on the active $|\mathfrak{q}_a\rangle$ and inert $|\mathfrak{q}_s\rangle$ elementary states.*

Any even-dimensional algebra $\mathcal{C}\ell_{p,q}$ over the field $\mathbb{F} = \mathbb{R}$ is isomorphic to the tensor product of two-dimensional algebras $\mathcal{C}\ell_{0,2}$ ($\mathbb{K} \simeq \mathbb{H}$) and $\mathcal{C}\ell_{1,1}$, $\mathcal{C}\ell_{2,0}$ ($\mathbb{K} \simeq \mathbb{R}$). Accordingly, we have ***two types of elementary states***

I. $|\mathfrak{q}_a\rangle = \left|\mathbb{H}, 0, 1, \frac{1}{2}\right\rangle$, $|\bar{\mathfrak{q}}_a\rangle = \left|\overline{\mathbb{H}}, 0, -1, \frac{1}{2}\right\rangle$ active state, and

II. $|\mathfrak{q}_s\rangle = \left|\mathbb{R}, 0, 0, \frac{1}{2}\right\rangle$, $|\bar{\mathfrak{q}}_s\rangle = |\mathfrak{q}_s\rangle$ inert state.

In terms of the binary structure, there are three basic operations on states with a minimal tensor dimension:
1) ***Fusion***.

$$|\mathfrak{q}\rangle \otimes |\bar{\mathfrak{q}}\rangle = \left|\mathbb{H}, 0, 1, \frac{1}{2}\right\rangle \otimes \left|\overline{\mathbb{H}}, 0, -1, \frac{1}{2}\right\rangle = |\mathbb{H} \otimes \overline{\mathbb{H}}, 0, 0, 1\rangle = |\mathbb{R}, 0, 0, 1\rangle = |\gamma\rangle .$$

2) ***Doubling***.

$$\begin{aligned} \left|e^-\right\rangle &= |\mathfrak{q}\rangle \oplus |\bar{\mathfrak{q}}\rangle = \left|\mathbb{H} \oplus i\mathbb{H}, 0, 1, \frac{1}{2}\right\rangle = \left|\mathbb{C}, 0, 1, \frac{1}{2}\right\rangle, \\ \left|e^+\right\rangle &= |\mathfrak{q}\rangle \ominus |\bar{\mathfrak{q}}\rangle = \left|\mathbb{H} \ominus i\mathbb{H}, 0, -1, \frac{1}{2}\right\rangle = \left|\overline{\mathbb{C}}, 0, -1, \frac{1}{2}\right\rangle . \end{aligned}$$

3) ***Annihilation***.

$$\begin{aligned} &\left|e^-\right\rangle \otimes \left|e^+\right\rangle = \left|\mathbb{C}, 0, 1, \frac{1}{2}\right\rangle \otimes \left|\overline{\mathbb{C}}, 0, -1, \frac{1}{2}\right\rangle = \\ &= \left|\mathbb{H} \oplus i\mathbb{H}, 0, 1, \frac{1}{2}\right\rangle \otimes \left|\mathbb{H} \ominus i\mathbb{H}, 0, -1, \frac{1}{2}\right\rangle = |\mathbb{H} \otimes \mathbb{H}, 0, 0, 1\rangle + i\,|\mathbb{H} \otimes \mathbb{H}, 0, 0, 1\rangle - \\ &\qquad - i\,|\mathbb{H} \otimes \mathbb{H}, 0, 0, 1\rangle + |\mathbb{H} \otimes \mathbb{H}, 0, 0, 1\rangle = 2\,|\mathbb{R}, 0, 0, 1\rangle = 2\,|\gamma\rangle . \end{aligned}$$

The latter operation is an algebraic analog of the electron-positron annihilation: $e^-e^+ \longrightarrow 2\gamma$. For more details, see [24].

According to Theorem 2, the *fermionic states* F correspond to cyclic vectors $\boldsymbol{\tau}_{\frac{k}{2},0}\otimes\boldsymbol{\tau}_{0,\frac{r}{2}}|\omega\rangle$ with an odd number of cofactors $\boldsymbol{\tau}_{\frac{1}{2},0}$ (resp. $\boldsymbol{\tau}_{0,\frac{1}{2}}$) in the tensor product. In turn, the *bosonic states* B correspond to cyclic vectors with an even number of cofactors. Therefore,

$$\underbrace{\boldsymbol{\tau}_{\frac{1}{2},0}\otimes\boldsymbol{\tau}_{\frac{1}{2},0}\otimes\cdots\otimes\boldsymbol{\tau}_{\frac{1}{2},0}\bigotimes\boldsymbol{\tau}_{0,\frac{1}{2}}\otimes\boldsymbol{\tau}_{0,\frac{1}{2}}\otimes\cdots\otimes\boldsymbol{\tau}_{0,\frac{1}{2}}}_{m\text{ times}}\Rightarrow\left\{\begin{array}{lll} F, & m\equiv 1 & \pmod 2; \\ B, & m\equiv 0 & \pmod 2. \end{array}\right. \tag{3}$$

It follows that the cyclic vector of the $\mathbb{K}$-Hilbert space, representing bosons or fermions depending on $m\equiv 0,1 \pmod 2$, has a separable structure.

3. The Concept of a Particle

The holistic approach also faces the problem of defining a particle. In contrast to the reductionist framework, where particles are principal elements and parts make up a whole, in the holistic Universe, parts are only subordinates under the whole. A part has no autonomous status but represents some state of the whole, a certain mode of its being. Thus, in the holistic approach, a particle is not an individual object but rather a state of substance (primary substrate), which is considered as the whole. In this regard, the problem arises of an alternative description of the main characteristics of particles in a holistic approach.

3.1. Mass

In the case of the fundamental symmetry $G_f = \mathrm{SO}_0(1,3)$ (Lorentz group), the set Ω of pure separable states ω on the algebra $\mathfrak{A}$ corresponds to a system of cyclic vectors of the form

$$\pi_\omega(\mathfrak{h}^{(1)})\pi_\omega(\mathfrak{h}^{(2)})\cdots\pi_\omega(\mathfrak{h}^{(n)}|\omega\rangle \longmapsto \boldsymbol{\tau}_{\frac{k}{2},0}\otimes\boldsymbol{\tau}_{0,\frac{r}{2}}|\omega\rangle \tag{4}$$

in a $\mathbb{K}$-Hilbert space $\mathbf{H}(\mathbb{K})$, where $\mathfrak{h}^{(i)}\in\mathfrak{A}$, $i=1,2,\ldots,n$; $\pi_\omega \simeq \boldsymbol{\tau}_{l\dot{l}}$ is the spinor representation of the group $\mathbf{Spin}_+(1,3)$, $\mathbb{K}=\mathbb{R},\mathbb{C},\mathbb{H}$. Physical $\mathbb{K}$-Hilbert space $\mathbf{H}_{\text{phys}}(\mathbb{K})$ (spectrum of matter (energy)) is formed by cyclic vectors (4), for which the mass of the pure separable state ω is determined by the formula [14]

$$\boxed{m_\omega = m_e\left(l+\tfrac{1}{2}\right)\left(\dot{l}+\tfrac{1}{2}\right)}$$

Here m_e is the rest mass of the electron. Thus, the mass (energy) of the states ω is given by the tensor structure of cyclic vectors $|\psi\rangle\in\mathbf{H}_{\text{phys}}(\mathbb{K})$.

3.2. Charge

The charge of the states ω is determined within the framework of the $\mathbb{K}$-linear structure of the space $\mathbf{H}_{\text{phys}}(\mathbb{K})$: $\mathbb{K} \simeq \mathbb{C}$ – charged states, $\mathbb{K} \simeq \mathbb{H}$ – neutral states, $\mathbb{K} \simeq \mathbb{R}$ – truly neutral states.

Theorem 3 ([24]) *Let the C^*-algebra $\mathfrak{A}$ consist of the energy operator H and the generators of the group $\mathrm{SU}(2,2)$ attached to H, forming a general system of eigenfunctions with H. And let the set of pure separable states ω on the algebra $\mathfrak{A}$ correspond to a system of cyclic vectors $|\psi\rangle$ in the GNS-Hilbert space H_ω. Then the $\mathbb{K}$-linear structure ($\mathbb{K} = \mathbb{R}, \mathbb{C}, \mathbb{H}$) translates H_ω into a physical $\mathbb{K}$-Hilbert space $\mathbf{H}_{\text{phys}}(\mathbb{K})$, in which three basis sectors are allocated:*
1) Charged state sector $\mathbf{H}_{\text{phys}}(\mathbb{C})$.
2) Neutral state sector $\mathbf{H}_{\text{phys}}(\mathbb{H})$.
3) Truly neutral state sector $\mathbf{H}_{\text{phys}}(\mathbb{R})$.

3.3. Spin

Unlike the mechanical definition of the spin of a quantum microobject ($q\bar{q}$-meson or qqq-baryon) of the quark model, in the algebraic formulation, we have a non-classical (group-theoretic) definition of this most important characteristic: $s = l - \dot{l}$. The entire spectrum of states (see the cone of representations in Fig 1 in [13], and also [23]) is divided into a sequence of spin lines along which the states have the same spin but different mass (tensor structure). The first theory providing a correct mathematical formulation of the "classically non-describable two-valuedness" of the electron spin was proposed by Pauli in 1927. Avoiding any visual mechanical models, Pauli introduced a doubled Hilbert space $\mathbf{H}_2 \otimes \mathbf{H}_\infty$ (vector space of wave functions), whose vectors are two-component spinors. That was the first advent of two-component spinors in physics and the first case of doubling. The next doubling that yielded the $\mathbf{H}_4 \otimes \mathbf{H}_\infty$ space of bispinors was accomplished by Dirac in 1928. The next-to-next doubling leads to the Hilbert space of hyper-twistors $\mathbf{H}_8 \otimes \mathbf{H}_\infty$ [20, 21]. Hyper-twistors are vectors of the fundamental representation of the Rumer-Fet group $\mathrm{SO}(2,4) \otimes \mathrm{SU}(2) \otimes \mathrm{SU}(2)'$, which provides a group-theoretic interpretation of the periodic system of elements [17]. Spinors, bispinors, and twistors are special cases of hyper-twistors.

Doubling (or duality) is a universal feature of matter. The spin of an electron is a manifestation of duality rather than an internal property of a thing. Recall that electrons can be observed only having a certain direction of spin and not both of them simultaneously. Duality preexists in matter (substance), which endows its accidences (electrons) with a certain value of spin depending on the experimental situation (manifestation).

3.4. Discrete Symmetries

Other important characteristics of a state (along with mass, charge, and spin) are discrete symmetries and associated quantum numbers (P-parity, C-parity, etc.). And here, unlike the mechanical definitions, which depend on the angular moments of the quark model, the algebraic approach is more universal and does not depend on any classical (macroscopic) definitions. Namely, the Clifford algebra $C\ell$ is associated with each cyclic vector (4). In the case of the number field $\mathbb{F} = \mathbb{C}$, eight automorphisms (including the identical automorphism Id) are defined for the Clifford algebra $\mathbb{C}_n$[4]. Let's list these transformations and their spinor representations:

$$
\begin{array}{lcll}
\mathcal{A} \longrightarrow \mathcal{A}^\star & : & \mathsf{A}^\star = \mathsf{W}\mathsf{A}\mathsf{W}^{-1}, & \\
\mathcal{A} \longrightarrow \widetilde{\mathcal{A}} & : & \widetilde{\mathsf{A}} = \mathsf{E}\mathsf{A}^{\mathsf{T}}\mathsf{E}^{-1}, & \\
\mathcal{A} \longrightarrow \widetilde{\mathcal{A}^\star} & : & \widetilde{\mathsf{A}^\star} = \mathsf{C}\mathsf{A}^{\mathsf{T}}\mathsf{C}^{-1}, & \mathsf{C} = \mathsf{E}\mathsf{W}, \\
\mathcal{A} \longrightarrow \overline{\mathcal{A}} & : & \overline{\mathsf{A}} = \Pi\mathsf{A}^{*}\Pi^{-1}, & \\
\mathcal{A} \longrightarrow \overline{\mathcal{A}^\star} & : & \overline{\mathsf{A}^\star} = \mathsf{K}\mathsf{A}^{*}\mathsf{K}^{-1}, & \mathsf{K} = \Pi\mathsf{W}, \\
\mathcal{A} \longrightarrow \overline{\widetilde{\mathcal{A}}} & : & \overline{\widetilde{\mathsf{A}}} = \mathsf{S}\left(\mathsf{A}^{\mathsf{T}}\right)^{*}\mathsf{S}^{-1}, & \mathsf{S} = \Pi\mathsf{E}, \\
\mathcal{A} \longrightarrow \overline{\widetilde{\mathcal{A}^\star}} & : & \overline{\widetilde{\mathsf{A}^\star}} = \mathsf{F}\left(\mathsf{A}^{*}\right)^{\mathsf{T}}\mathsf{F}^{-1}, & \mathsf{F} = \Pi\mathsf{C}.
\end{array}
$$

Here, the symbol T stands for transposition, $*$ for complex conjugation. W, E, Π are matrices of the automorphisms $\mathcal{A} \to \mathcal{A}^\star$, $\mathcal{A} \to \widetilde{\mathcal{A}}$, $\mathcal{A} \to \overline{\mathcal{A}}$ in the spinor representation. It is easy to verify that the set of automorphisms $\{\mathrm{Id}, \star, \widetilde{\ }, \widetilde{\star}, \overline{\ }, \overline{\star}, \overline{\widetilde{\ }}, \overline{\widetilde{\star}}\}$ forms a finite group of the eighth order. This group, by virtue of the commutativity $\widetilde{(\mathcal{A}^\star)} = \left(\widetilde{\mathcal{A}}\right)^\star$, $\overline{(\mathcal{A}^\star)} = \left(\overline{\mathcal{A}}\right)^\star$, $\overline{\left(\widetilde{\mathcal{A}}\right)} = \widetilde{\left(\overline{\mathcal{A}}\right)}$, $\overline{\left(\widetilde{\mathcal{A}^\star}\right)} = \widetilde{\left(\overline{\mathcal{A}}\right)}^\star$ and the involution property $\star\star = \widetilde{\ }\widetilde{\ } = \overline{\ }\,\overline{\ } = \mathrm{Id}$, is isomorphic to the cyclic group $\mathbb{Z}_2 \otimes \mathbb{Z}_2 \otimes \mathbb{Z}_2$ (the Cayley table of this group is shown in Table 1). Next, let $\mathrm{Ext}(\mathbb{C}_n) = \{\mathrm{Id}, \star, \widetilde{\ }, \widetilde{\star}, \overline{\ }, \overline{\star}, \overline{\widetilde{\ }}, \overline{\widetilde{\star}}\}$ be the automorphism group of the algebra $\mathbb{C}_n$, then there is an isomorphism be-

[4] In 1955, Rashevsky [25] showed that there are *four fundamental automorphisms* of the algebra $\mathbb{C}_n$: $\mathcal{A} \to \mathcal{A}$ (identity), $\mathcal{A} \to \mathcal{A}^\star$ (involution), $\mathcal{A} \to \widetilde{\mathcal{A}}$ (reversion) and $\mathcal{A} \to \widetilde{\mathcal{A}^\star}$ (conjugation), where $\mathcal{A}$ is an arbitrary element of the algebra $\mathbb{C}_n$. The group structure of the set of automorphisms $\{\mathrm{Id}, \star, \widetilde{\ }, \widetilde{\star}\}$ with respect to discrete transformations that make up the PT group (the so-called *reflection group*) was studied in [26, 6]. Along with the fundamental automorphisms of the algebra $\mathbb{C}_n$ there is a *pseudo-automorphism* $\mathcal{A} \to \overline{\mathcal{A}}$ (for more details, see [25]), which is not fundamental, but its composition with the fundamental automorphisms allows us to extend the set $\{\mathrm{Id}, \star, \widetilde{\ }, \widetilde{\star}\}$ by means of pseudo-automorphisms $\mathcal{A} \to \overline{\mathcal{A}}$, $\mathcal{A} \to \overline{\mathcal{A}^\star}$, $\mathcal{A} \to \overline{\widetilde{\mathcal{A}}}$, $\mathcal{A} \to \overline{\widetilde{\mathcal{A}^\star}}$. The group structure of the *extended set of automorphisms*, $\{\mathrm{Id}, \star, \widetilde{\ }, \widetilde{\star}, \overline{\ }, \overline{\star}, \overline{\widetilde{\ }}, \overline{\widetilde{\star}}\}$, was studied in [27, 7] in relation to CPT symmetries.

Table 1: Cayley table of the group of involutive automorphisms of the algebra $\mathbb{C}_n$.

	Id	$\star$	$\sim$	$\widetilde{\star}$	$-$	$\overline{\star}$	$\overline{\sim}$	$\overline{\widetilde{\star}}$
Id	Id	$\star$	$\sim$	$\widetilde{\star}$	$-$	$\overline{\star}$	$\overline{\sim}$	$\overline{\widetilde{\star}}$
$\star$	$\star$	Id	$\widetilde{\star}$	$\sim$	$\overline{\star}$	$-$	$\overline{\widetilde{\star}}$	$\overline{\sim}$
$\sim$	$\sim$	$\overline{\star}$	Id	$\star$	$\overline{\sim}$	$\overline{\widetilde{\star}}$	$-$	$\overline{\star}$
$\widetilde{\star}$	$\widetilde{\star}$	$\sim$	$\star$	Id	$\overline{\widetilde{\star}}$	$\overline{\sim}$	$\overline{\star}$	$-$
$-$	$-$	$\overline{\star}$	$\overline{\sim}$	$\overline{\widetilde{\star}}$	Id	$\star$	$\sim$	$\widetilde{\star}$
$\overline{\star}$	$\overline{\star}$	$-$	$\overline{\widetilde{\star}}$	$\overline{\sim}$	$\star$	Id	$\widetilde{\star}$	$\sim$
$\overline{\sim}$	$\overline{\sim}$	$\overline{\widetilde{\star}}$	$-$	$\overline{\star}$	$\sim$	$\widetilde{\star}$	Id	$\star$
$\overline{\widetilde{\star}}$	$\overline{\widetilde{\star}}$	$\overline{\sim}$	$\overline{\star}$	$-$	$\widetilde{\star}$	$\sim$	$\star$	Id

tween $\mathrm{Ext}(\mathbb{C}_n)$ and the CPT group of discrete transformations[5]: $\mathrm{Ext}(\mathbb{C}_n) \simeq \{1, P, T, PT, C, CP, CT, CPT\} \simeq \mathbb{Z}_2 \otimes \mathbb{Z}_2 \otimes \mathbb{Z}_2$. At this point, the inversion of space P, the time reversal T, the complete reflection PT, the charge conjuga-tion C, the transformations CP, CT and the complete CPT transformation correspond to the automorphism $\mathcal{A} \rightarrow \mathcal{A}^\star$, the anti-automorphisms $\mathcal{A} \rightarrow \widetilde{\mathcal{A}}$, $\mathcal{A} \rightarrow \widetilde{\mathcal{A}^\star}$, the pseudo-automorphisms $\mathcal{A} \rightarrow \overline{\mathcal{A}}$, $\mathcal{A} \rightarrow \overline{\mathcal{A}^\star}$, the pseudo-anti-automorphisms $\mathcal{A} \rightarrow \overline{\widetilde{\mathcal{A}}}$ and $\mathcal{A} \rightarrow \overline{\widetilde{\mathcal{A}^\star}}$, respectively [27, 7].

In general, all elements of the group $\mathrm{Ext}(\mathbb{C}_n)$ depend on phase factors [23, 10]. Let

$$P = \eta_p \mathsf{W}, \quad T = \eta_t \mathsf{E}, \quad C = \eta_c \Pi,$$

where $\eta_p, \eta_t, \eta_c \in \mathbb{C}^* = \mathbb{C} - \{0\}$ are the phase factors. Then for $\mathbb{F} = \mathbb{C}$ we have

$$\begin{aligned} \mathrm{Ext}(\mathbb{C}_n) &\simeq \{1, P, T, PT, C, CP, CT, CPT\} \simeq \\ &\simeq \{\mathbf{1}_{n/2}, \eta_p \mathsf{W}, \eta_t \mathsf{E}, \eta_p \eta_t \mathsf{EW}, \eta_c \Pi, \eta_c \eta_p \Pi \mathsf{W}, \eta_c \eta_t \Pi \mathsf{E}, \eta_c \eta_p \eta_t \Pi \mathsf{EW}\} \simeq \\ &\simeq \{\mathbf{1}_{n/2}, \eta_p \mathsf{W}, \eta_t \mathsf{E}, \eta_p \eta_t \mathsf{C}, \eta_c \Pi, \eta_c \eta_p \mathsf{K}, \eta_c \eta_t \mathsf{S}, \eta_c \eta_p \eta_t \mathsf{F}\}. \end{aligned} \tag{5}$$

In the case of a number field $\mathbb{F} = \mathbb{R}$, we have groups $\mathrm{Ext}(C\kern-0.2em\ell_{p,q})$ of the form (5), where the algebras $C\kern-0.2em\ell_{p,q}$ have a quaternion division ring $\mathbb{K} \simeq \mathbb{H}$ ($p - q \equiv 4, 6 \pmod 8$) or a real ring $\mathbb{K} \simeq \mathbb{R}$ ($p - q \equiv 0, 2 \pmod 8$). Corresponding phase factors: $\eta_p, \eta_t, \eta_c \in \mathbb{H}^* = \mathbb{H} - \{0\}$, $\eta_p, \eta_t, \eta_c \in \mathbb{R}^* = \mathbb{R} - \{0\}$.

[5]It is interesting to note that in the Penrose twistor program [28], the spinor structure is understood as the underlying (more fundamental, primary) structure with respect to Minkowski space-time. In other words, the space-time continuum is not a fundamental substance in the twistor approach. The continuum is an absolutely derived entity (in the spirit of Leibnitz's relational philosophy) generated by the underlying spinor structure. In this context, the space-time discrete symmetries P and T are projections (shadows) of the fundamental automorphisms of the spinor structure.

4. Non-locality

According to Axiom **A.I**, at the fundamental level, all states of a single quantum system **U** form a common system of eigenfunctions. In this case, all states of the system **U** are connected (pass into each other) under the action of the fundamental symmetry group G_f. Due to the generality of the definition of the system **U** and the flexibility of the GNS construction, for each specific implementation of the operator algebra (the so-called "dressing" of the C^*-algebra), we obtain our own (corresponding to this implementation) spectrum of states of the system **U**. So, in the case when the generators of the fundamental symmetry group attached to H ($G_f = \mathrm{SO}_0(1.3)$ – Lorentz group) are generators of the group algebra $\mathfrak{sl}(2,\mathbb{C})$, we obtain a linearly growing mass spectrum of states ("elementary particles"). In this case, the various states of the system **U** are connected to each other by means of Weyl generators (ladder operators of the algebra $\mathfrak{sl}(2,\mathbb{C})$, for more details, see [29]). In the case of a periodic system of elements (fundamental symmetry $G_f = \mathrm{SO}(4,2)$ is a conformal group), twelve Weyl generators of the group algebra $\mathfrak{su}(2,2)$ perform transitions between elements (transmutation of elements).

According to Axiom **A.I** and Theorem 1, the principle of superposition is valid for the states of a single quantum system **U** forming a $\mathbb{K}$-Hilbert space. This means that combinations of fermionic and bosonic states (3)-(4) of the **U** (matter spectrum) system form both separable and non-separable (entangled) states, which leads to the effect of nonlocality.

As is known [30], an algebraic bipartition of the C^*-algebra $\mathfrak{A}$ is any pair $(\mathfrak{A}_1, \mathfrak{A}_2)$ of subalgebras $\mathfrak{A}_1, \mathfrak{A}_2 \subset \mathfrak{A}$ such that $\mathfrak{A}_1 \cap \mathfrak{A}_2 = 1_{\mathfrak{A}}$. This directly implies the concept of operator locality.

Definition 1 *An element of an algebra $\mathfrak{A}$ is called local with respect to a given bipartition $(\mathfrak{A}_1, \mathfrak{A}_2)$ or $(\mathfrak{A}_1, \mathfrak{A}_2)$-local if this element is the product $\mathfrak{a}_1\mathfrak{a}_2$ of the element $\mathfrak{a}_1 \in \mathfrak{A}_1$ and the element $\mathfrak{a}_2 \in \mathfrak{A}_2$.*

The following definition establishes the most important concepts of separability and entanglement of states on algebra $\mathfrak{A}$.

Definition 2 *A state ω on the algebra $\mathfrak{A}$ is called separable with respect to a bipartition $(\mathfrak{A}_1, \mathfrak{A}_2)$ if the expectation $\omega(\mathfrak{a}_1\mathfrak{a}_2)$ of any local operator $\mathfrak{a}_1\mathfrak{a}_2$ can be decomposed into a linear convex combination of products of expectations*

$$\omega(\mathfrak{a}_1\mathfrak{a}_2) = \sum_k \lambda_k \omega_k^{(1)}(\mathfrak{a}_1)\omega_k^{(2)}(\mathfrak{a}_2), \quad \lambda_k \geq 0, \ \sum_k \lambda_k = 1, \tag{6}$$

where $\omega_k^{(1)}$ and $\omega_k^{(2)}$ are states on the algebra $\mathfrak{A}$. Otherwise, the state ω is called ***entangled*** *with respect to the bipartition $(\mathfrak{A}_1, \mathfrak{A}_2)$.*

This definition of separability can be easily extended to the case of more than two partitions. For example, for the case of n-partition we have

$$\omega(\mathfrak{a}_1\mathfrak{a}_2\cdots\mathfrak{a}_n)=\sum_k\lambda_k\omega_k^{(1)}(\mathfrak{a}_1)\omega_k^{(2)}(\mathfrak{a}_2)\cdots\omega_k^{(n)}(\mathfrak{a}_n),\quad \lambda_k\geq 0,\ \sum_k\lambda_k=1.$$

When the state ω is pure, the separability condition (6) is simplified.

Definition 3 *Pure states ω on the operator algebra $\mathfrak{A}$ are separable with respect to a given bipartition $(\mathfrak{A}_1,\mathfrak{A}_2)$ only and if only*

$$\omega(\mathfrak{a}_1\mathfrak{a}_2)=\omega(\mathfrak{a}_1)\omega(\mathfrak{a}_2)$$

for all local operators $\mathfrak{a}_1\mathfrak{a}_2$.

Hence, pure separable states are product states. Taking into account the GNS construction and Axiom **A.III**, the general form of any pure separable state is given by the following theorem.

Theorem 4 *Let the state ω on the algebra $\mathfrak{A}$ be separable with respect to a given bipartition $(\mathfrak{A}_1,\mathfrak{A}_2)$. Then the normed pure state $|\psi\rangle$ in a physical $\mathbb{K}$-Hilbert space $\mathbf{H}_{\rm phys}(\mathbb{K})$ is $(\mathfrak{A}_1,\mathfrak{A}_2)$-separable if and only if*

$$|\psi\rangle=\pi_\omega(\mathfrak{b}^{(1)})\pi_\omega(\mathfrak{b}^{(2)})\,|\omega\rangle\,,$$

where $\mathfrak{b}^{(i)}\in\mathfrak{A}$, $i=1,2$, $\pi_\omega(\mathfrak{b}^{(i)})$ is a cyclic representation of the algebra $\mathfrak{A}$ in the $\mathbb{K}$-Hilbert space $\mathbf{H}_{\rm phys}(\mathbb{K})$.

It is obvious that for the case of algebraic n-partition, the pure separable state has the form

$$|\psi\rangle=\pi_\omega(\mathfrak{b}^{(1)})\pi_\omega(\mathfrak{b}^{(2)})\cdots\pi_\omega(\mathfrak{b}^{(n)})\,|\omega\rangle\,.$$

By virtue of the correspondence $\omega\leftrightarrow|\psi\rangle$ (Axiom **A.III**) for cyclic vectors $|\psi\rangle\in\mathbf{H}_{\rm phys}(\mathbb{K})$, there are such combinations of the form (6) for which the condition of definition 2 is not fulfilled, i.e., the state determined by such a combination is entangled.

Conclusion

In conclusion, it should be noted that one of the main goals of this article was the desire to emphasize the fundamental role of two-component spinors. A two-component spinor describes the minimal structural component of matter (a quantum of energy). Identification of the minimal structural component $|\mathfrak{q}\rangle$ (quantum of energy) with neutrino $|\nu\rangle$, $|\mathfrak{q}\rangle \equiv |\nu\rangle$, would lead to the "neutrino theory of everything" and to the understanding of neutrino as the primary element of matter. However, the substance (energy) is not determined by its states. As Spinoza said: "Substance is by nature the first of its states" (Theorem 1, Ethics). Substance as a whole is more primary than its states. States have a secondary (subordinate) nature in relation to the whole (substance). This is one of the main principles of holism. The spectrum of states of the substance, which we call energy or matter, contains an infinite number of states, among which one of the most minimal is the neutrino. Substance is not an aggregate constructed from elementary parts, contrary to all the ideas of atomism and reductionism. The quantized nature of the Fermi and Bose states of the matter spectrum consists in the factorization (separability) of cyclic vectors of the $\mathbb{K}$-Hilbert space by means of the tensor product of fundamental states $|\mathfrak{q}\rangle$ (GNS construction, algebraic quantization). At the same time, the presence of a $\mathbb{K}$-structure splits $|\mathfrak{q}\rangle$ into two states $|\mathfrak{q}_a\rangle$ and $|\mathfrak{q}_s\rangle$.

References

[1] Dirac, P.A.M.: *Lectures on Quantum Field Theory*. Yeshiva University, New York (1967).

[2] Bohm, D.: *Wholeness and the Implicate Order*. Routledge, London-New York (2002).

[3] Sakharov, A.D.: Is there an elementary length? In *Collection of scientific works*, Tsentrkom, Moscow, Russia, 1995, pp. 384–397. (in Russian).

[4] Varlamov, V.V.: On the system of axioms of nonlocal quantum theory. *Mathematical structures and modeling* **4(44)**, 5–25 (2017). (in Russian)

[5] von Neumann, J.: *Mathematical Foundations of Quantum Mechanics*. Princeton University Press, NJ (1955).

[6] Varlamov, V.V.: Discrete Symmetries and Clifford Algebras. *Int. J. Theor. Phys.* **40**, 769–805 (2001).

[7] Varlamov, V.V.: Universal Coverings of Orthogonal Groups. *Adv. Appl. Clifford Algebr.* **14**, 81–168 (2004).

[8] Varlamov, V.V.: CPT groups for spinor field in de Sitter space. *Phys. Lett. B.* **631**, 187–191 (2005).

[9] Varlamov, V.V.: CPT Groups of Higher Spin Fields. *Int. J. Theor. Phys.* **51**, 1453–1481 (2012).

[10] Varlamov, V.V.: CPT groups of spinor fields in de Sitter and anti-de Sitter spaces. *Adv. Appl. Clifford Algebras* **25**, 487–516 (2015).

[11] Dyson, F.: The threefold way: algebraic structure of symmetry groups and ensembles in quantum mechanics. *J. Math. Phys.* **3**, 1199–1215 (1962).

[12] Baez, J.C.: Division Algebras and Quantum Mechanics. *Found. Phys.* **42**, 819–855 (2012).

[13] Varlamov, V.V.: Algebraic quantum mechanics I.: Basic definitions. *Mathematical structures and modeling.* **2(54)**, 4–23 (2020).

[14] Varlamov, V.V.: Mass quantization and the Lorentz group. *Mathematical structures and modeling.* **2(42)**, 11–28 (2017).

[15] Varlamov, V.V.: Lorentz Group and Mass Spectrum of Elementary Particles. arXiv:1705.02227 (2017).

[16] Workman, R.L.: *et. al.* Particle Data Group. *Prog. Theor. Exp. Phys.* 083C01 (2022).

[17] Fet, A.I.: *Group Theory of Chemical Elements.* Gruyter, Berlin-Boston (2016).

[18] Varlamov, V.V.: Group-theoretic description of the periodic system of elements. *Mathematical structures and modeling.* **2(46)**, 5–23 (2018).

[19] Varlamov, V.V.: Group-theoretic description of the periodic system of elements II: Seaborg table. *Mathematical structures and modeling.* **1(49)**, 5–21 (2019).

[20] Varlamov, V.V.: Group-theoretic description of the periodic system of elements III: 10-periodic extension. *Mathematical structures and modeling.* **3(51)**, 5–20 (2019).

[21] Varlamov, V.V., Pavlova, L.D, Babushkina, O.S.: Group Theoretical Description of the Periodic System. *Symmetry.* **14**, 137 (2022).

[22] Barut, A., Raczka, R.: *Theory of Group Representations and Applications*, PWN, Warsczawa (1977).

[23] Varlamov, V.V.: Spinor Structure and Internal Symmetries. *Int. J. Theor. Phys.* **54**,10, 3533–3576 (2015).

[24] Varlamov, V.V.: Spinors in $\mathbb{K}$-Hilbert Spaces. arXiv:2204.10808 (2022).

[25] Rashevskii P.K.: The Theory of Spinors. *Uspekhi Mat. Nauk.* **10**, 3–110 (1955); English translation in Amer. Math. Soc. Transl. (Ser. 2). **6**, 1 (1957).

[26] Varlamov V.V.: Fundamental Automorphisms of Clifford Algebras and an Extension of Dąbrowski Pin Groups. *Hadronic J.* **22**, 497–535 (1999).

[27] Varlamov, V.V.: *Group Theoretical Interpretation of the CPT-theorem.* Mathematical Physics Research at the Cutting Edge (Ed. C. V. Benton) New York, Nova Science Publishers (2004) P. 51–100; arXiv:math-ph/0306034 (2003).

[28] Penrose, R.: The twistor programme. *Rep. Math. Phys.* **12**, 65–76 (1977).

[29] Varlamov, V.V.: Group Theory and Mass Quantization. arXiv:2311.16175 (2023).

[30] Benatti, F., Floreanini, R.: Entanglement in Algebraic Quantum Mechanics: Majorana fermion systems. *J. Phys. A: Math. Theor.* **49**, 305303 (2016).

Chapter 8

Kinematics and Relativistic Mechanics on the Basis of Appell Polynomials

R. M. Yamaleev *
Joint Institute for Nuclear Research, Laboratory of Information Technology, Dubna, Russia

Abstract

The ladder structure of kinematics and the associated dynamical equations arising from time-dependent forces represented by polynomial functions is examined. Unifying all orders of acceleration into a single sequence of Appell polynomials establishes an analogy between kinetic energy and momentum. This approach makes the formulation of extended kinematics similar to traditional mechanics and restores the notion of a phase space. Within this unified framework, the relativistic mass-shell equation is extended to models characterized by higher-degree characteristic polynomials. Finally, a superstructure composition for relativistic mechanics is proposed.

Keywords: kinematics, evolution equations, Pascal matrix, mass-shell equation, Appell polynomials

1. Introduction

The polynomial representation of dynamical variables is a widely used technique in mathematical calculations. This work applies this mathematical tool to kinematics,

---

*Corresponding Author's Email: yamaleev@jinr.ru

In: Mathematical Problems in Relativity, Gravitation, and Cosmology
Editor: Valeriy Dvoeglazov
ISBN: 979-8-89530-622-2

where the fundamental elements of the theory are represented as sequences of polynomials. We develop generalized kinematic framework in which distance, velocity, acceleration, and higher-order accelerations are determined using higher-degree polynomials. The evolution equations derived within this framework generally take the form of triangular systems composed of Appell sequences of polynomials and Pascal matrices. [1, 2]. Starting from relatively simple mathematical considerations, it will be shown that such a realization involves only Appell polynomials and Pascal matrices. In mathematics an Appell sequence is any polynomial sequence $\{p_n(x)\}_{n=0,1,2,\ldots}$, satisfying the identity

$$\frac{d}{dx}p_n(x) = np_{n-1}(x),$$

and in which $p_0(x)$ is a non-zero constant. The characteristic polynomials of generalized classical relativistic mechanics also form an Appell series of polynomials which can be treated as an extension of Hermite polynomials [3].

In this paper we will consider the classical aspects of the kinematics higher- order accelerations and we will treat the problem *ab initio*. Just to give an idea of the problem we are going to treat, we consider some similar features of the kinematics of inertial motion and accelerated motion with constant acceleration.

Within the framework of the developed mathematical tool, the relativistic mass-shell polynomial is extended to the Appell sequence of polynomials. Boundary conditions corresponding to typical physical scenarios yield quantization of the factor-coefficients that connect the set of invariant values of the evolution equations with a unique quantityâĂŤthe proper mass of the particle. In this paper, we will examine classical aspects of the kinematics of higher-order accelerations and approach the problem *ab initio*. To illustrate the nature of the problem we intend to address, we highlight certain analogous features between the kinematics of inertial motion and motion under constant acceleration.

Inertial motion refers to movement with uniform, constant velocity. The displacement associated with such motion as a function of time is represented by a linear polynomial in time:

$$S(t + t_0) = vt + S(t_0). \tag{1.1}$$

Conventionally it is assumed that the motions are divided into two classes: (I) inertial motions, (II) non-inertial motions.

One difference between these two types of motions is that in inertial motion, the main characteristic, velocity, is a constant, while the main characteristic of non-inertial motion, acceleration, is generally a function of the time. However, let us observe that motion with uniform constant acceleration exhibits features similar to inertial motion. Uniform motion with constant acceleration refers to motion in which the velocity of an object changes by equal amounts over equal intervals of time. The displacement traveled by a particle under a constant force is expressed as a quadratic polynomial function of time.

$$s(t+t_0) = \frac{a}{2}t^2 + v(t_0)t + s(t_0), \tag{1.2}$$

where $s(t_0)$ and $v(t_0)$ mean the initial data, and a is the constant acceleration. Let two distinct real numbers t_1, t_2 be defined by the following function of the initial data:

$$v(t_0) = \frac{a}{2}(t_1+t_2), s(t_0) = \frac{a}{2}(t_1 t_2). \tag{1.3}$$

Then, the following holds true

$$s(t+t_0) = \frac{a}{2}(t+t_1)(t+t_2). \tag{1.4}$$

It can be seen that the polynomial $s(t)$ is obtained by shifting the pair of quantities t_1, t_2âĂŃ. It should also be noted that, under this shift, the second coefficient of the polynomial (1.2) changes according to the equation:

$$v(t+t_0) = at + v(t_0). \tag{1.5}$$

This is a universal rule.

In other words, the distance is constructed by extending the product of the roots of the polynomial. Clearly, this pair of roots is defined by the initial motion data $s(t_0)$ and $v(t_0)$. Thus, within this representation, the acceleration can be described by the action of a shift operator that shifts the pair of initial points t_1, t_2 along the time parameter t.

Under this translation the difference of roots

$$t_1 - t_2 = 2T, \tag{1.6}$$

remains unchanged. In terms of kinematics, the formula

$$T^2 = v(t_0)^2/a^2 - 2s(t_0)/a \tag{1.7}$$

is invariant under shifts of the initial time point,

$$v(t_0+\delta)^2/a^2 - 2s(t_0+\delta)/a = v(t_0)^2/a^2 - 2S(t_0)/a = T^2. \tag{1.8}$$

Initial and final points are denoted by $t_0 + \delta = t_b, \;\; t_0 = t_a$. Then,

$$\frac{1}{2a}(v^2(t_b) - v^2(t_a)) = s(t_b) - s(t_a). \tag{1.9}$$

In these terms, the kinetic energy is defined as

$$E_{kin} = \frac{mv^2}{2} - \frac{ma^2}{2}T^2. \tag{1.10}$$

2. Peculiarities of Kinematics of Higher Order Accelerated Motion

In the equation (1.2) the acceleration a is a constant value and does not change during accelerated motion. The change in speed over time is defined by the first order polynomial

$$v(t+t_0) = at + v(t_0). \tag{2.1}$$

In the case of variable acceleration the following degrees of expansion should be included. The polynomial of third degree refers to motion with a constant *jerk*. The distance equation is given by the polynomial of the third degree

$$s(t+t_0) = \frac{j}{6}t^3 + \frac{1}{2}a(t_0)t^2 + v(t_0)t + s(t_0). \tag{2.2}$$

The coefficient at the highest degree, j, is called *jerk*. This polynomial also can be viewed as the result of displacement of the initial data,

$$s(t+t_0) = \frac{j}{6}(t+t_1)(t+t_2)(t+t_3), \tag{2.3}$$

where (t_1, t_2, t_3) are defined as follows

$$s(t_0) = \frac{j}{6}t_1t_2t_3,\ v(t_0) = \frac{j}{6}(t_1t_2 + t_2t_3 + t_3t_1),\ a(t_0) = \frac{j}{3}(t_1 + t_2 + t_3).$$

The equation (2.3) depends on three initial data, they change in the process of motion, that means, it is necessary to add two more evolution equations for the coefficients, these are,

$$v(t+t_0) = \frac{j}{2}t^2 + a(t_0)t + v(t_0), \tag{2.4}$$

$$a(t+t_0) = jt + a(t_0). \tag{2.5}$$

The system of three equations, (2.2), (2.4), (2.5), describes the evolution of three initial values: the distance, speed and the acceleration. A further generalization of this scheme is obvious. In order to identify a certain pattern, let us consider the case when the distance traveled is determined by a fourth degree polynomial:

$$s(t+t_0) = \frac{y}{24}t^4 + \frac{1}{6}j(t_0)t^3 + \frac{1}{2}a(t_0)t^2 + v(t_0)t + s(t_0), \tag{2.6}$$

$$v(t+t_0) = \frac{y}{6}t^3 + \frac{1}{2}j(t_0)t^2 + a(t_0)t + v(t_0),$$

$$a(t+t_0) = \frac{y}{2}t^2 + j(t_0)t + a(t_0),$$

$$\jmath(t+t_0) = yt + \jmath(t_0). \tag{2.7}$$

Such sequences of polynomials can be represented compactly using the shift matrix θ^+. The entries of this matrix are 1âĂŹs along the sub-diagonal and 0âĂŹs elsewhere:

$$\theta^+_{ij}(n) = \left(\begin{array}{cc} 1, & \text{if } i = j+1 \\ 0, & \text{otherwise,} \end{array} \right) \quad i,j = 0,1,2,...,n. \tag{2.8}$$

In addition, let us define a state vector

$$[S(t);n] = (\ldots, \jmath(t), a(t), v(t), s(t))^T. \tag{2.9}$$

Within these notations the system of equations is presented as follows

$$[S(t+t_0);5] = \exp(t\theta^+(5))[S(t_0);5]. \tag{2.10}$$

In this representation, the transition to the normal form of the polynomials is achieved through a change of variables in the shift matrix. Utilizing the equation

$$\jmath(t+t_0) = yt + \jmath(t_0), \tag{2.11}$$

in (2.10) replace the time variable with the difference

$$t = \frac{1}{y}(\jmath(t+t_0) - \jmath(t_0)). \tag{2.12}$$

As a consequence, equation (2.10) is transformed as follows

$$[S(t+t_0);5] = \exp(\frac{1}{y}(\jmath(t+t_0) - \jmath(t_0))\theta^+(5))[S(t_0);5], \tag{2.13}$$

or,

$$\exp(-\frac{1}{y}\jmath(t+t_0))\theta^+(5))[S(t+t_0)] = \exp(-\frac{1}{y}\jmath(t_0))\theta^+(5))[S(t_0)] = [R(5)]. \tag{2.14}$$

From this equation, it follows that the vector $[R(5)]$ is a vector with invariant entries. For this reason, in general, we will refer to this vector as an *invariant vector* and denote it by

$$[R(n)] = \left(\begin{array}{ccccc} 1 & 0 & r_2 & \cdots & r_n \end{array} \right)^T \tag{2.15}$$

In order to obtain the set of polynomials in terms of the invariant coefficients we must apply the following formula

$$[S(\jmath/y);n] = \exp(\frac{\jmath}{y}\theta^+(n))[R(n)]. \tag{2.16}$$

For $n = 5$, we get

$$\begin{pmatrix} y \\ j \\ a(j) \\ v(j) \\ s(j) \end{pmatrix} = \begin{pmatrix} 1 & 0 & 0 & 0 & 0 \\ \frac{j}{y} & 1 & 0 & 0 & 0 \\ \frac{1}{2}(\frac{j}{y})^2 & \frac{j}{y} & 1 & 0 & 0 \\ \frac{1}{6}(\frac{j}{y})^3 & \frac{1}{2}(\frac{j}{y})^2 & \frac{j}{y} & 1 & 0 \\ \frac{1}{24}(\frac{j}{y})^4 & \frac{1}{6}(\frac{j}{y})^3 & \frac{1}{2}(\frac{j}{y})^2 & \frac{j}{y} & 1 \end{pmatrix} \begin{pmatrix} y \\ 0 \\ r_2 \\ r_3 \\ r_4 \end{pmatrix} \tag{2.17}$$

It is important to emphasize that the formulas for the invariant make use of the concept of the work performed by the top force $f_y = my$. For convenience, let us adopt units where $m = 1$. Let us demonstrate that the invariants r_2, r_3, r_4 are obtained as constants of integration along the directions a, v, s, respectively. The *work* done by y along the coordinate $[a]$ is defined by the integral:

$$\int_1^2 yda = \int_1^2 \frac{dj}{dt}da = \int_1^2 jdj. \tag{2.18}$$

Form these equations it follows

$$a = \frac{1}{2y}j^2 + r_2. \tag{2.19}$$

Definition of the *work* done by y along the coordinate $[v]$:

$$\int y\, dv = \int \frac{dj}{dt}dv = \int (\frac{1}{2y}j^2 + r_2)dv$$

gives

$$v = \frac{1}{6y^2}j^3 + \frac{r_2}{y}j + r_3. \tag{2.20}$$

It can be seen that the formulas for $a(j), v(j)$ obtained from the matrix equation (2.17) coincide with the formulas (2.19), (2.20).

This kind of higher- acceleration motion is generated by the force defined as polynomials of the time. For example, let us define force as a quadratic polynomial of the form

$$F_N = \frac{1}{2}F_y t^2 + F_j t + F_a. \tag{2.21}$$

According to Newtonian equation

$$F_N = \frac{d}{dt}(mv) = ma. \tag{2.22}$$

For the next derivative of the acceleration we have

$$\frac{d}{dt}F_N = \frac{d}{dt}(ma) = mj = F_y t + F_j. \tag{2.23}$$

Finally, we get

$$\frac{d^2}{dt^2}F_N = \frac{d}{dt}(mj) = my = F_y. \tag{2.24}$$

Integrating the system of equations of motion (2.21)-(2.24) we shall arrive at the system of equations of fourth degree.

Before closing this section we note that for fourth degree, $n = 4$, we can define the formula for the distance as the kinetic energy

$$E_{kin}(j/y) = \frac{y}{24}(\frac{j}{y})^4 + r_2\frac{1}{6}(\frac{j}{y})^2 + r_3\frac{j}{y} + r_4.$$

In this context, the quantity $p = \frac{j}{y}$ plays the role of momentum. Then,

$$v = \frac{ds}{dp}$$

The functional dependence of velocity on momentum is given by the formula. These formulas encourage us to interpret the expressions of extended kinematics using the terminology of classical mechanics.

3. Comparison of the System of Higher-order Equations with Hermite Polynomials

The discussion in the previous section can be easily generalized, and the procedure for constructing the system of invariants is straightforward. The solution of the Newtonian equations under a force defined by a truncated Taylor expansion of order $(n-2)$ is presented by the following set of equations:

$$f(t) = \frac{f^{(n)}}{n!}t^n + \frac{f^{(n-1)}}{(n-1)!}t^{n-1} + \frac{f^{(n-2)}}{(n-2)!}t^{n-2} + \cdots + f(t_0), \tag{3.1}$$

$$f^{(1)}(t) = f^{(1)}(t) \cdots f^{(k)}(t) = f^{(k)}(t),$$

$$\cdots$$

$$f^{(n-1)}(t) = f^{(n)}(t_0) + f^{(n-1)}(t_0). \tag{3.2}$$

Redefining coefficients

$$p_k(t) = k!\frac{f^{(k)}(t)}{f^{(n)}},\ k = 1, 2, 3, \ldots, n. \tag{3.3}$$

Introducing the state vector

$$[P(t); n] = (1, p_1(t), \cdots, p_n(t))^T. \tag{3.4}$$

Defining the following nilpotent matrix

$$A^{sub} = (k-1)\delta_{k,k-1}, k = 1, \ldots, n. \tag{3.5}$$

Then, the sequences of polynomials (3.1)-(3.2) can be cast in the matrix form as follows

$$[P(t+t_0); n] = \exp(tA^{sub})[P(t_0); n], \tag{3.6}$$

where $\exp(tA^{sub})$ is the Pascal matrix.

In order to obtain the set of polynomials with invariant coefficients, the time variable must be replaced by a difference:

$$t = p_1(t) - p_1(t_0), \tag{3.7}$$

in equation(3.6). We get

$$[R] = \exp(-p_1(t)A^{sub})[P(p_1(t)); n] = \exp(-p_1(t_0)A^{sub})[P(p_1(t_0)); n]. \tag{3.8}$$

Hence the resulting vector

$$[R] = \begin{pmatrix} 1 & 0 & r_2 & \cdots & r_n \end{pmatrix}^T \tag{3.9}$$

which can be referred as the *invariant vector*. From formulae (3.8)-(3.9) we arrive at the system of polynomials with invariant coefficients

$$\begin{gathered} p_1 = p_0 \\ p_2 = p_0^2 + r_2 \\ p_3 = p_0^3 + 3r_2p_0 + r_3 \\ p_4 = p_0^4 + 6r_2p_0^2 + 4r_3p_0 + r_4 \\ p_5 = p_0^5 + 10r_2p_0^3 + 10r_3p_0^2 + 5r_4p_0 + r_5 \\ \ldots \\ p_n = p_0^n + C_n^2r_2p_0^{n-2} + \cdots + C_n^kr_kp_0^{n-k} + \cdot + C_n^1r_{n-1}p_0 + r_n \end{gathered} \tag{3.10}$$

It may be useful to compare this system of equations with the set of Hermite polynomials [5]

$$\begin{gathered} He_1 = p_0 \\ He_2 = p_0^2 - 1 \\ He_3 = p_0^3 - 3p_0 \\ He_4 = p_0^4 - 6p_0^2 + 3 \\ He_5 = p_0^5 - 10p_0^3 + 15p_0 \\ He_6 = p_0^6 - 15p_0^4 + 45p_0^2 - 15 \end{gathered} \tag{3.11}$$

As can be seen, in these equations $r_2 = -1, r_3 = 0, r_4 = 3, r_5 = 0, r_6 = -15$. In the same manner as demonstrated above, the system (3.11) can be expressed in the following matrix form:

$$[P(p)] = \exp(p_0 A(n))[R]. \tag{3.12}$$

4. Extension of the Relativistic Mass-shell Equation to Higher Degree Characteristic Polynomials

One advantage of the present formulation is the possibility of using this mathematical tool to extend relativistic kinematics. First of all, let us note that upon examination of the second equation of the polynomial system (3.11), a natural parallel emerges with the equation

$$p_2 = p_0^2 + r_2, \tag{4.1}$$

and the relativistic mass-shell equation

$$r_2 = -m^2, (c = 1). \tag{4.2}$$

Firstly, let us demonstrate that evolution in relativistic mechanics can be expressed in terms of the Pascal matrix representation. Consider a quadratic polynomial of the form

$$p^2(u) = p^2(0) + 2up_0 + u^2, \tag{4.3}$$

$$p_0(u) = p_0 + u. \tag{4.4}$$

In terms of the $(3x3)$ matrix we get

$$\begin{pmatrix} 1 \\ p_0(u) \\ p^2(u) \end{pmatrix} = \begin{pmatrix} 1 & 0 & 0 \\ u & 1 & 0 \\ u^2 & u & 1 \end{pmatrix} \begin{pmatrix} 1 \\ p_0(0) \\ p^2(0) \end{pmatrix} \tag{4.5}$$

Using the same algorithm, let us determine the invariant functions:

$$\begin{pmatrix} 1 \\ 0 \\ -m^2 \end{pmatrix} = \begin{pmatrix} 1 & 0 & 0 \\ -p_0 & 1 & 0 \\ p_0^2 & -2p_0 & 1 \end{pmatrix} \begin{pmatrix} 1 \\ p_0 \\ p^2 \end{pmatrix} \tag{4.6}$$

We find

$$p^2 = p_0^2 - m^2,$$

with $(c = 1)$.

Let us define the components of the hyper-kinetic energy by the column vector:

$$[P(u)] = \begin{pmatrix} 1 & p_1(u) & p_2^2(u) & \cdots & p_n^n(u) \end{pmatrix}^T \tag{4.7}$$

Then, the evolution of the hyper-kinetic vector is expressed by the following equation

$$[P(u)] = \exp(uA(n))[P(0)] \tag{4.8}$$

We will derive the invariant values of this evolution process by replacing the parameters. Substituting here $u = p_1(u) - p_1(0)$, the resulting equation can be expressed as follows:

$$\exp(-p(u)A(n))[P(u)] = \exp(-p(0)A(n))[P(0)] = [R]. \tag{4.9}$$

Hence the resulting vector is

$$[R] = \begin{pmatrix} 1 & 0 & r_2 & \cdots & r_n \end{pmatrix}^T \tag{4.10}$$

which has been denoted as the *invariant vector.* From formulae (4.7)-(4.10) we arrive at the system of polynomials with invariant coefficients

$$\begin{aligned} p_1 &= p \\ p_2^2 &= p^2 - m^2 \\ p_3^3 &= p^3 - 3pm^2 + r_3 \\ p_4^4 &= p^4 - 6p^2m^2 + 4r_3p + r_4 \\ p_5^5 &= p^5 - 10p^3m^2 + 10r_3p^2 + 5r_4p + r_5 \\ p_6^6 &= p^6 - 15p^4m^2 + 20r_3p^3 + 15r_4p^2 + 6r_5p + r_6 \\ p_7^7 &= p^7 - 21p^5m^2 + 35r_3p^4 + 35r_4p^3 + 21r_5p^2 + 7r_6p + r_7 \\ &\cdots \end{aligned} \tag{4.11}$$

In matrix form, this system can be represented by the following equation:

$$[P(p)] = \exp(p_oA(n))[R]. \tag{4.12}$$

As can be seen, the second equation of this system of polynomials can be identified with the mass-shell equation.

Let us outline the procedure leading to extension of the relativistic connection between mass and the kinetic energy. It is convenient to introduce a pivot polynomial as follows. Define product of two quantities [6],

$$p_2^2(0) = q_1q_2,$$

and perform a shift of the quantities using the indefinite variable. We then obtain a quadratic polynomial of the form:

$$p_2^2(u) = (q_1^2 + u)(q_2^2 + u) = u^2 + 2p_o(0)u + p_2^2(0). \tag{4.13}$$

In addition, we get the equation

$$p_o(u) = u + p_o(0), \tag{4.14}$$

where

$$p_0(0) = \frac{1}{2}(q_1^2 + q_2^2), \tag{4.15}$$

If we interpret p_2^2 as the squared momentum, $p_2^2 = p^2$, then we arrive at relativistic mechanics. Transforming equation (4.15) into a polynomial with invariant coefficients yields the mass-shell equation:

$$p^2 = p_0^2 - r^2,$$

where the invariant is defined by the equation

$$p^2(u) - p_0^2(u) = p^2(0) - p_0^2(0) = -r^2 = -(mc)^2. \tag{4.16}$$

The velocity is defined in the usual manner

$$\frac{dp_0}{dp} = \frac{V}{c} = \frac{p}{p_0}. \tag{4.17}$$

Let us follow the same procedure for the case of a third-degree polynomial. Define the pivot polynomial as a shift of the roots of the free coefficient [7]:

$$p_3^3(0) = q_1 q_2 q_3.$$

By prolongation of the roots we get

$$\begin{aligned} p_3^3(u) &= (q_1 + u)(q_2 + u)(q_3 + u) = \\ &= u^3 + 3p_1(0)u^2 + 3p_2^2(0)u + p_3^3(0). \end{aligned} \tag{4.18}$$

This equation is a result of the prolongation of the free coefficient. For the other coefficients, as for,

$$\begin{aligned} p_1(0) &= \frac{1}{3}(q_1 + q_2 + q_3), \\ p_2^2(0) &= \frac{1}{3}(q_1 q_2 + q_2 q_3 + q_3 q_1), \end{aligned} \tag{4.19}$$

we obtain an additional equation,

$$p_2^2(u) = u^2 + 2p_1(0)u^2 + p_2(0), \tag{4.20}$$

$$p_1(u) = u + p_1(0).$$

In invariant coefficients we get

$$\begin{aligned} p_1 &= p \\ p_2^2 &= p^2 + r_2 \\ p_3^3 &= p^3 + 3r_2 p + r_3 \end{aligned} \tag{4.21}$$

One may consider this system of equations as analogous to (or an extension of) the mass-shell equation [8].

Similarly, by analogy, we define two velocities by

$$\frac{dp}{dp_2} = \frac{p_2}{p}, \text{(in units} c = 1) \tag{4.22}$$

and,

$$\frac{dp}{dp_3} = \frac{p_3^2}{p_2^2} = V_3, (c = 1). \tag{4.23}$$

In this context, the variable p should be interpreted as p_0. Additionally, following this scheme, we have to define the k-th velocity by

$$\frac{dp}{dp_k} = \frac{p_k^{k-1}}{p_{k-1}^{k-1}} = V_k, k = 2, \cdots, n.(c = 1) \tag{4.24}$$

Let us define the following factor-coefficients

$$k_j = \frac{r_j}{m^j}, \ j = 2, 3, \dots \tag{4.25}$$

and denote the fraction by $m/p = d$. Then, the system of equations is modified as follows

$$\begin{array}{c} p_2^2/p^2 = 1 - d^2 \\ p_3^3/p^3 = 1 - 3d + k_3 d^3 \\ p_4^4/p^4 = 1 - 6d^2 + 4k_3 d^3 + k_4 d^4 \\ p_5^5/p^5 = 1 - 10d^2 + 10k_3 d^3 + 5k_4 d^4 + k_5 d^5 \\ p_6^6/p^6 = 1 - 15d^2 + 20k_3 d^3 + 15k_4 d^4 + 6k_5 d^5 + k_6 d^6 \\ p_7^7/p^7 = 1 - 21d^2 + 35k_3 d^3 + 35d_4 d^4 + 21k_5 d^5 + 7k_6 d^6 + k_7 d^7 \\ \cdots \end{array} \tag{4.26}$$

Consider two boundary conditions. The first is,

$$\frac{p_k^k}{p^k} = 1$$

This condition is satisfied by setting $d = 0$. Physically, this corresponds to the relativistic case of a massless particle, characterized by $m = 0$ and $V = c$.

The second condition corresponds to the rest state, characterized by $m \geq 0$ and $V = 0$, i.e., zero velocity. Additionally, the requirement that all momentum components, represented by variables on the left-hand side of equation (4.26), simultaneously vanish leads to a specific definition of the factor-coefficients $k_j, j = 3, \dots, n$. These coefficients can be determined recursively by the formula:

$$k_j = (j - 1)^{j-1}, \ j = 3, \cdots, n. \tag{4.27}$$

Correspondingly, the polynomials in equation (4.11) take the compact form

$$p_{j+1} = (x-1)^j(x+j), j = 1, 2, 3 \ldots.$$

Returning to the initial designations, we write

$$p_{j+1} = (p-m)^j(p+jm), \ j = 1, 2, 3, \ldots. \tag{4.28}$$

It is worth emphasizing that the mass spectrum coincides with the commutation relation of the matrices $A(n)$ and $\theta(n)^-$ reported in (3.5) and (2.8).

Finally, we arrive at the following series of the polynomials

$$\begin{gathered} g_2 = x^2 - 1 \\ g_3 = x^3 - 3x + 2 \\ g_4 = x^4 - 6x^2 + 8x - 3 \\ g_5 = x^5 - 10x^3 + 20x^2 - 15x + 4 \\ g_6 = x^6 - 15x^4 + 40x^3 - 45x^2 + 24x - 5 \\ g_7 = x^7 - 21x^5 + 70x^4 - 105x^3 + 84x^2 + 35x + 6 \\ \ldots \end{gathered} \tag{4.29}$$

This result can be realized as quantization of the mass parameter under boundary condition

$$p_j = 0 \text{ with } d = 1,$$

which implies that in the rest state all components of the momentum must be equal to zero. We thus conclude that we have received a *layered structure*. On the first level we have

$$p_2^2 = (p_0 - m)(p_0 + m), (c = 1). \tag{4.30}$$

On the second level the characteristic equation is given by the third degree polynomial

$$p_3^3 = (p_0 - m)(p_0 - m)(p_0 + 2m), \ (c = 1). \tag{4.31}$$

On the k-th level, the characteristic equation is defined as

$$p_k^k = (p_0 - m)^{k-1}(p_0 + (k-1)m), \ (c = 1). \tag{4.32}$$

Note that, within this classical framework, the mass spectrum is represented by a column vector whose components are invariant. Thus, an extension of classical mechanics is achieved, defined by the system of characteristic equations given previously (4.11).

Let γ be a Lorentz factor,

$$\gamma = \frac{1}{\sqrt{1 - V^2/c^2}}.$$

Then, from eq.(4.32) we obtain

$$\frac{p_k^k}{m^k} = (\gamma - 1)^{k-1}(\gamma + (k-1)), \ (c = 1), \ k = 2, 3, \ldots. \tag{4.33}$$

This formula expresses components of the momentum as functions of a single parameter- the velocity.

Conclusion

The technique of polynomial representation outlined here leads naturally to a layered structure, in which higher-order kinematics acquires a ladder-like organization. Although the ladder structure emerges without explicitly introducing external ladder operators, the resulting system of evolution equations inherently displays multiple levels organized according to the degree of polynomials that form the Appell sequence.

The method outlined here is fundamentally general and straightforward for practical applications. Its usefulness has been previously demonstrated in transforming truncated Taylor series into Appell polynomial sequences.

Specifically, the second-degree polynomial equation in the sequences obtained can be identified with the relativistic mass-shell equation. This observation prompts us to use the Appell sequences as the superstructure of relativistic mechanics.

The Appell polynomial sequences can be viewed as a generalization of Hermite polynomials, bringing the current theory closer to finite-dimensional quantum mechanics. Additionally, it is important to note that the split decomposition of a continuous acceleration function can be interpreted independently of external forces, highlighting its intrinsic role in higher-order kinematics.

Acknowledgment

The author is indebted to Dr. Jan Busa for helpful discussions and technical assistance.

References

[1] Aceto L., Trigiante D. The Matrices of Pascal and Classical Polynomials. *Rendicoti del Circoi Matematico in Palermo* (2002) Ser II. Suppl. 68: 219–228.

[2] Aceto L., Trigiante D. The Matrices of Pascal and Other Greats. *Amer. Math. Monthly* (2001) 108: 232–245.

[3] Yamaleev R.M. Evolution Equations of Higher Order Kinematics. *JINR Communications* (2023) Dubna P5-2023-54: 1–15

[4] Yamaleev R.M. Pascal Matrix Representation of Evolution of Polynomials. *Int. J. Appl. Comput. Math.* (2015) I(4): 513–525.

[5] Courant, Richard; Hilbert, David (1989) [1953], Methods of Mathematical Physics, vol. 1, Wiley-Interscience, ISBN 978-0-471-50447-4

[6] Yamaleev R.M. Generalized Newtonian Equations of Motion. *Ann. Phys.* (1999) 277: 1–18.

[7] Yamaleev R.M. Relativistic Equations of Motion within Nambu' s Formalism of Dynamics. *Ann. Phys.* (2000) 285: 141–160.

[8] Yamaleev R.M. Generalized Lorentz-force Equations. *Ann. Phys.* (2001) 292: 157–178.

Index

G

H

I

K

L

M

N

P

Q

R

S

T

U

V

W